BUSINESS AND TECHNICAL WRITING

An annotated bibliography of books, 1880-1980

Gerald J. ALRED
Diana C. REEP
Mohan R. LIMAYE

With the assistance of
MICHAEL A. MIKOLAJCZAK

The Scarecrow Press, Inc.
Metuchen, N.J., & London
1981

Library of Congress Cataloging in Publication Data

Alred, Gerald J
    Business and technical writing.

        Includes indexes.
        1.   Business report writing--Bibliography.   2.   Technical
writing--Bibliography.   I.   Reep, Diana C. , joint author.
II.   Limaye, Mohan R. , joint author.   III.   Title.
Z7164. C81A413   [HF5719]   016. 808'0666021   80-29211
ISBN 0-8108-1397-1

To our_____

families and friends_____

for their patience_____

CONTENTS_____

This bibliography is the result of a three-year effort to examine and annotate all books dealing with business or technical writing published within the last century. We considered a hundred years as the reasonable period encompassing what most would call the modern era of business and technical writing. Accordingly, we did not include in our study such things as sixteenth-century English letter-writing guides or examples of technical writing dating to the Roman descriptions of aqueducts.

In order to deal reasonably with the volume of material, we established two major limits. First, we limited our listings to book-length works or works published in book format. Hence, we do not include journal articles or papers read at conferences if they were not published separately. The second limit was that the books we listed had to deal significantly with writing or the analysis of writing, either for business or in technical and professional contexts. We therefore did not include books on rhetoric or the art of writing in general. Likewise, we rejected books that did not deal primarily with writing even though their titles implied that they did.

We gathered these titles from a variety of sources, including all existing bibliographies, lists in journals and books, reviews, and Books in Print. The next step was to find a copy of each book. We personally searched four major research libraries. We called for hundreds of titles through interlibrary-loan services. We wrote to publishers. We appealed to colleagues for advice. One discovery we made was the plethora of inaccuracies in the citations we had collected. Often authors' names were misspelled; publishing information was incorrect; and sometimes no such book existed. We made every effort to be comprehensive, but we cannot state beyond all doubt that we have missed nothing.

This bibliography is composed of five major divisions: an introduction, an annotated list of bibliographies, an annotated list of books, supplementary lists, and three indexes.

The Introduction is an overview of the works in the bibliography. The second division is an annotated list of all bibliographies of significant length. However, we have not included reference lists in textbooks even though these lists may be lengthy. The bibliographies we have annotated vary widely in style, completeness, and focus. Some cover only the collection in a particular library; some cover subjects that we do not include in our own annotated lists. We include these bibliographies as further sources for the reader, particularly in areas we do not include here.

The third division is the annotated list of books. The information in each annotation comes from our direct examination of the book. The bibliographic citations include information as shown on the title pages. If the citation includes "Index," the book contains an index to the entire contents. If the citation includes "Bib.," the book contains a substantial and separate reference section that was not incorporated into chapters. The number of pages cited refers to Arabic-numbered pages.

The annotations themselves contain our assessment of the purpose of the book, its scope, primary and unusual topics covered, pedagogical materials, and historical interest. In most cases, we annotated the most recent edition that we could obtain. Although new editions invariably contain new material, reflect new organization, or offer new pedagogical devices, we believed that it was more useful to annotate an earlier edition when we could not find the latest than to cite the latest without annotation. However, in the case of books published before 1935, we tried to find the earliest edition, since we believed that the historical value of such works is paramount.

Some entries in both the bibliography and book lists are not annotated. Those are references to works that, in spite of our best efforts, we could not find. We include the references because they appeared in other lists. But we must emphasize that since we have not seen these books, we cannot vouch for the accuracy of the citation itself or for whether the book belongs in this bibliography.

The fourth division consists of supplementary lists of books on several subjects. Many of these, such as military style guides, are related to our main topic because they are works often needed by those in business and technical writing. These books are cited as they appeared in other compilations, and these lists are not exhaustive.

The fifth division consists of the three indexes--by coauthor, by title, and by subject--that the reader will use to find an annotation of a particular book. These indexes do not include the bibliographies and books on supplementary lists.

We hope that this bibliography will offer both historical and practical information to instructors, researchers, and students in business and technical writing.

Our deepest thanks go to all those who lent support to this project and contributed to its success: Father Terence McCloskey and Ruth Schimke of the library of Holy Redeemer College, Waterford, Wisconsin, and Jessica Brown and the Library staff of the University of Wisconsin/Milwaukee for devoting incredible effort to tracking down books; Professor L. W. Denton, Auburn University; Professor Francis W. Weeks, University of Illinois; Professor Kathryn Whitford, Jerre Collins, Pamela Johnson, and Elizabeth Larsen of the University of Wisconsin/Milwaukee--all of whom offered valuable suggestions; and Shirley Dement, who typed the manuscript. We must especially acknowledge the assistance of Michael A. Mikolajczak, of Marquette University, without whom this project would not have been completed within three years.

Gerald J. Alred
The University of Wisconsin--Milwaukee

Diana C. Reep
The University of Akron

Mohan R. Limaye
The University of Texas at Austin

> The letter is the great business builder of the present decade. With its extended use is coming a corresponding improvement in the style of writing--a style that is free from meaningless formality and full of life, clear and strong in its appeal to the reader. --Ion E. Dwyer, The Business Letter (Boston: Houghton Mifflin, 1914), preface.

> It has been said that in this age the man of science appears to be the only one who has anything to say, and he is the one that least knows how to say it. --T. A. Rickard, A Guide to Technical Writing (San Francisco: Scientific and Mining Press, 1908), p. 7.

As we examined business and technical writing books written over the last hundred years, we discovered that the turn of the century ushered in a burst of publication, especially in business writing. Two developments may be significant in causing this rise in publication. The first is the development of business instruction, when R. M. Bartlett in Philadelphia in 1846 began teaching bookkeeping and other commercial subjects. By 1860, all major cities had business instruction, and by 1900, business instruction in colleges and universities was not uncommon. Second, the modern typewriter was placed on the market in 1874 by E. Remington and Son. By 1900, the typewriter (both the person and the machine) was commonplace in business. The effect that these two influences--business instruction and the typewriter--had on the genesis of business writing guides is a potential area for scholarship and study.

One of the earliest titles on correspondence for the professional in business is the first edition of Mercantile Correspondence in English and German Languages, by John Clausen (ca. 1889). This work was designed to train business people in how to read and write letters in German and English. Another early work is Hand-Book of Modern Business Correspondence (1908), by Forrest Crissey. This work was designed to "guide ... students, stenographers, correspondence clerks, all office men, and managers" in the art of letter

writing in business. Calvin Althouse, in an early work titled Business
Letters (1910), asserted that "one of the great needs of the business
world today is to find those who can indite a good business letter,
cordial in tone, correctly punctuated and expressed in clear, unmis-
takable English. . . . " One might hear a similar statement today--
over seventy years later.

By 1916, a number of books attempted to answer Althouse's
call with general treatments of business correspondence (Poole and
Buzzell, 1913) and specific treatments of sales letters (Rose, 1915),
of direct-mail letters (Ruxton, 1918, and Sales Promotion by Mail,
1916), and of house organs (Wilson, 1915).

The publication of textbooks on business writing is directly
related to business writing instruction. According to Professor
Francis W. Weeks, the first courses in business writing were offered at the University of Illinois (1902) and New York University
(1903). Originally published in 1904, Commercial Correspondence
and Postal Information, by Altmaier, may rank as the first business
writing textbook. It contains not only writing assignments but also
cases for study. LaSalle Extension University published Business
English (1911), by Edwin Herbert Lewis, for students from high
school through college. Another business writing textbook, by H. W.
Hammond, in its fifth edition by 1913, was designed for schools where
no teacher of business English was employed, and it served as a supplement to regular business courses.

Edwin Hall Gardner and Robert Ray Aurner's Effective Business Letters, originally published in 1915, was based on a business-
letter writing course at the University of Wisconsin. Nathaniel
Waring Barnes's How to Teach Business Correspondence (1916) might
be considered the earliest work on pedagogy. However, it could
also be considered a teacher's manual for his How to Write Business
Letters (1916). By 1923, two books were exclusively devoted to business writing courses in high schools: Handbook for Business Letter
Writers (1922) and Actual Business English (1923).

Technical Writing

With a singular exception, early works on technical writing were tied
to the education of engineers and scientists. Sir T. Clifford Allbutt's Notes on the Composition of Scientific Papers, originally published in 1904, served as a guide for medical students in preparing
theses. As far as we can determine, the first technical writing
book aimed strictly at a professional audience is Thomas Arthur
Rickard's A Guide to Technical Writing (1908). This work, aimed
at metallurgical engineers and geologists, has a modern flavor
throughout, especially with its emphasis on audience. Its author,
the editor of Scientific and Mining Press, includes a paper he delivered to the American Association for the Advancement of Science
at Denver on August 28, 1901, titled "A Plea for Greater Simplicity
in the Language of Science. "

Although Ray Palmer Baker's The Preparation of Reports

(1924) has been called the first technical writing textbook, * there
are at least three earlier texts.   The Elements of Specification Writing:
A Textbook for Students in Civil Engineering, originally published in
1913, was designed for both the student and the practicing engineer.
This book served as a writing, as well as a legal, guide for the
preparation of construction specifications.   Homer Watt in The Composition of Technical Papers, first published in 1917, gives advice
on writing business letters as well as technical papers.   Karl Owen
Thompson's Technical Exposition (1922) is a general technical
writing text, based on a course at Case School of Applied Science.
Thompson points out that his work is designed "to cover the more
practical of the instruction in English which follow the groundwork
in composition and rhetoric. " Baker, perhaps better than Thompson,
provides a clearer definition of technical writing when he suggests
that "a report must be so built as to carry the right view of a matter perhaps highly technical in nature to readers who have no first-
hand knowledge of the case. "   Both Thompson's and Baker's remarks
illustrate a common assumption of both business and technical
writing texts, namely, that pragmatic writing should follow a course
in general composition.

Pedagogy

Not until the middle of this century did works exclusively devoted to
pedagogy appear.   And not until the 1970s were there collections of
articles about the teaching of business and technical writing (Cunningham and Estrin, 1975; Sawyer, 1977; and Douglas, 1978).   In the
early 1960s, two books by James M. Brown, Casebook for Technical
Writers (1961) and Cases in Business Communication (1962), provided
materials for the classroom.   Shortly after, Stuart Levine's Materials
for Technical Writing (1963) provided both cases and exercises for
technical writing courses.   More recently, ABCA publications--
Business Communication Casebook (1974) and Business Communication Casebook Two (1977)--and Business Writing Cases and Problems
(1977), by Weeks and Hatch, provide help for instructors.
     Pedagogy of business writing recently has included texts based
on the creation and conduct of a fictitious company (Lindauer, 1974;
Melrose, 1977; Franco and Zall, 1978).   As Leonard Franco and
Paul M. Zall state, this approach gives the reader "a sense of what
it is like to write in a business-industrial environment, with one eye
on budgets and the other on schedules, your mind on the problem at
hand, and your emotions tied up in the act of writing. "
     Other pedagogical approaches have included home-study courses
(Campbell, 1976), programmed books (Reid and Silleck, 1978), boxed
materials (Reed, 1978), and the use of cartoons (Coffin, 1975).

---

*Richard W. Schmelzer, "The First Textbook on Technical Writing, "
The Journal of Technical Writing and Communication, 7 (1977), pp.
51-54.   Schmelzer refers to The Preparation of Reports, by Ray
Palmer Baker, as the first textbook.

Professional Technical Writing

Allan Lytel points out in Technical Writing as a Profession (1959)
that technical writing has its professional genesis in World War II.
During that time, writers were hired to produce a myriad of publications related to the maintenance, construction, and operation of highly
technical equipment.   After the war and during the heyday of the space
program, technical writers were hired by industry to produce documentation for increasingly complex systems.   Technical writers and
industrial publishing companies continue to link industry, business,
government, and the public.   Godfrey and Parr in The Technical
Writer (1959) discuss the similar development of the technical writer
in England.

     Malden Bishop's Go Write, Young Man (1961) defines the na-
ture of the profession and recommends that the best way for the
technical writer to get respect is to be a professional.   Bishop's
later book, Billions for Confusion: The Technical Writing Industry
(1963), looks critically at the profession.   Two books in the 1970s
(Clarke and Root, 1972; and Gould and Losano, 1975) provided career
guidance for entering the field.

     Numerous books have been written for professional technical
writers.   By far the most extensive work is the two-volume Handbook
of Technical Writing Practices (1971), an anthology of articles on a
wide range of subjects, edited by Stello Jordan et al.   Clifford Baker's Technical Publications: Their Purpose, Preparation and Production (1955) is a similar work.   Several other general writing
guides have been aimed at the professional (Hicks, 1959; Turner,
1964; Strong and Eidson, 1971).   Other books, such as Campbell
and Farrar's Effective Communication for the Technical Man (1972),
have been designed for technical writers in particular industries
(petroleum in this case).

     Some books focus on forms of technical writing, such as tech-
nical manuals (Walton, 1968), government proposals (Mandel and
Caldwell, 1962), and specifications (Sawyer, 1960).   Emerson Clarke
in A Guide to Technical Literature Production (1961) provides a guide
for managers of technical-publications departments, including a writer's aptitude test.   John Bernard Bennett's Editing for Engineers
(1970) suggests ways for editors to work tactfully with engineers.
Ehrlich and Murphy (1964) and Tichy (1966) also offer advice for
those who supervise writers.

Special Subjects

Some authors focus on individual disciplines and the writing and communication problems or strategies they require.   Gensler and Gensler
(1961), for example, discuss at length how chemists should keep a
laboratory notebook so as to allow them to incorporate their notes
into reports and papers.   Harris (1976) treats the special problems
in writing for the social sciences (e.g., multiple words for the same
concept).   Books by Cochran, Fenner, and Hill (1973) and Arny and
Reaske (1972) deal with writing for the earth sciences and ecology.
Grinsell, Rahtz, and Williams (1974) treat writing reports for British

archaeologists. Richardson and Callahan (1962) discuss the writing needs of home economists.

Other writers have concentrated on special fields and forms-- computer science (Smith, 1976), chemical engineering (Kobe, 1957), business English for non-native speakers (Kruse and Kruse, 1976), policies and instructions (Cooper, 1960), and scholarly writing (Mc- Cartney, 1953). Sherman Kent, in one such work, Writing History (1967), says, "Historical writing must have ... a continuous flow of clearly stated ideas. In this respect it differs most essentially from, say, the writing of belle lettres...."

Medical and legal writing have always been important, yet the number of works published is relatively limited. One well-known treatment is Morris Fishbein's Medical Writing: The Technic and the Art (1972). Report-Writing in Dentistry--A Teaching Outline (1960), by James Avery et al., has, through nine editions, been a popular guide for students, but more recently Communication in Den- tistry (1974), by Kenneth A. Easlick et al., has aimed at both stu- dents and practitioners. Hellen Morrin's Communication for Nurses (1959) treats writing for nurses in America; Austen and Crosfield in English for Nurses (1976) provide a similar, more recent, treatment for nurses in England.

For legal writing, one book has been designed for pre-law students (Brand and White, 1976) and another book for law students and practitioners (Cooper, 1963). Arthur M. Smith (1958) deals with the special problems of patent writing. Another book (Wattles, 1976) provides a very specialized treatment of writing legal descriptions of land.

Government prose, often a subject discussed within textbooks and other works, has been discussed in James R. Masterman and Wendell Brooks Phillips's Federal Prose: How to Write in and/or for Washington (1948) and Jefferson Bates's Writing with Precision: How to Write So That You Cannot Possibly Be Misunderstood; Zero Base Gobbledygook (1978). Both books, as their titles suggest, handle the subject with humor, even irony. Jud Monroe in Effective Re- search and Report Writing in Government (1980) provides special help in a more serious treatment. A related book, Letter Writing for Public Officials (1977), is aimed at state officials.

Reports have been treated generally in some works, e.g., David M. Robinson's Writing Reports for Management Decisions (1969). Other works have focused on specific types of reports-- most noticeably police reports (Gabard and Kenny, 1957; Patterson, 1977), accounting reports (Palen, 1955; Lewis, 1957; Joplin and Pat- tillo, 1969), and annual reports (Floyd, 1960; Koestler, 1969).

Letter Writing

Not surprisingly, numerous works have been devoted both to letters in general and to specific types of letters. By far, the most prolific authors of letter-writing books are William H. Butterfield, Lester Eugene Frailey, and J. Harold Janis. Some letter writing guides contain primarily models, such as Frailey's Handbook of Business Letters (1965). A subcategory of model-letter books are guides for

translating letters (Bar, 1971). These guides often show letters
side-by-side in four languages. Numerous works from early in this
century were devoted to direct-mail letters. Perhaps the most exhaustive modern treatment is Richard S. Hodgson's 1,575-page The
Dartnell Direct Mail Order Handbook (1974).

Special works have been devoted to sales letters, from J. C.
Aspley's Salesman's Correspondence Manual, 2nd ed. (1917), and
Charles R. Raymond's Modern Business Writing (1921). Other works
have dealt with "semibusiness" letters (Clapp, 1935). More recently,
Lassor A. Blumenthal in The Complete Book of Personal Letter-
Writing and Modern Correspondence (1969) combines business with
semibusiness correspondence, such as "Tenant-Landlord Letters."

Even more specialized book-length treatments are given to
adjustment letters (Prout, 1954) and credit and collection letters
(Tregoe and Whyte, 1924; Butterfield, 1941; Frailey, 1941). James
Tregoe and John Whyte discuss in Effective Collection Letters (1924)
the "you attitude" and the "stunt and humorous" collection letter.
Other works are devoted to letter writing for specific types of businesses, such as hotels and restaurants (Morrison and Montgomery,
1959) and banks (Butterfield, 1946).

Approaches to Writing

The works in this bibliography reveal that for the most part business
writing has blended psychology, communication theory, and business
practice with rhetoric and writing principles. Likewise, technical
writing has blended principles of objectivity, technical forms, audience analysis, and nonverbal elements with these rhetoric and writing
principles. Although some authors have stressed single approaches,
there has been considerable cross-fertilization between business and
technical writing.

Whatever their approaches, the authors represented in this
bibliography have recognized that they were teaching more than form
and ritual. Alexander Candee in Business Letter Writing (1920)
states that he does not want simply to teach letter forms but "correct mental principles" to "up-build for the good of the sender."
Robert Tuttle and C. A. Brown in Writing Useful Reports: Principles and Applications (1956) assert that they have rejected the
"types," "style," and "freshman composition" approaches in favor
of one in which "the report writer analyzes the report situation ...
examines the means [functional elements] at his disposal ... and
selects or modifies to fit the situation." And Thomas Johnson
in Analytical Writing: A Handbook for Business and Technical
Writers (1966) recognizes as a problem the "catalogical" approach
to writing, in which the writer simply lists his or her ideas.
He advocates an analytical approach, in which the writer analyzes
and then interprets data for the reader. Other more specialized
studies examine special problems in certain fields. Jack T. Huber in Report Writing in Psychology and Psychiatry (1961) examines the troubling conflict between usefulness and objectivity in psychological reports. John Dirckx (1977) explores the process of medical writing using analogy, as the titles of his major sections reveal:

"The Anatomy of Language" (which covers linguistics and sources of medicine's language) and "Sentences and Paragraphs: The Physiology of Language. "

A number of authors of scientific-writing books have been particularly interested in the analogous relationship of the scientific method to the writing process. Donald Menzel, Howard Mumford Jones, and Lyle G. Boyd in Writing a Technical Paper (1961) examine at length the composition process from the "evolution" of a first draft of a technical paper to its conclusion. Martin S. Peterson in Scientific Thinking and Scientific Writing (1961) suggests that the scientific method can be directly related to the writing process; for example, inductive thinking is translated into inductive paragraphs, and hypotheses become topic sentences. Judson Monroe, Carole Meredith, and Kathleen Fisher in The Science of Scientific Writing (1977) see a relationship between structure (format and development) and the reader's understanding, based on behavioral responses. In Writing Scientific Papers and Reports (1976), W. Paul Jones links the cognitive process inherent in the scientific method to the process of writing scientific papers and reports.

Many business and technical writing books have, in varying degrees, recognized a writing process in several steps. Since this process is recursive, however, a satisfactorily descriptive model has not yet emerged. In recent years, authors have sought to apply heuristics to the writing process. J. C. Mathes and Dwight W. Stevenson in Designing Technical Reports (1976), for example, provide the student with a procedure for determining audience needs and report design. Norman Tallent in Psychological Report Writing (1976) states: "The theme runs throughout the book that the psychologist, rather than presenting 'results' of his tests, interacts with his data and generates conclusions. . . . " He sees the writing process not as a one-dimensional fixed procedure but as a recursive, evolution of idea interacting with data. Tallent's work suggests the value of examining the body of information this bibliography represents: how scholars in the future can build new models of the processes of business and technical writing.

The works in this bibliography also reflect ways in which language, protocol, rhetorical strategies, and pedagogical devices have changed and have affected present-day practices and views. Consider the implications of Almonte C. Howell's advice to the writer in Military Correspondence and Reports (1943): "The writer should . . . not use the first personal pronoun . . . not even the second should be used. " Scholars might study the extent to which this advice has affected the current practice of business and technical writing. Models of business writing from the past also reveal social attitudes, as the following letter from The Business Letter-Writer's Manual (1924), by Charles Buck, reveals:

Dear Hayden:

With real regret I must write that it will not be possible for me to attend the reception on Friday evening. Important as is the obligation, I feel that there's one of equal importance to me--that of visiting with my little son.

Junior, who is now at Groton, has been in the infirmary for over a week. You know how my life is wrapped up in the boy, and since I have promised to spend the weekend with him, I must leave here on Friday afternoon for school.

I know you will understand, old man.

Sincerely yours,

Of course, the author could not anticipate the humor readers might find almost sixty years later--but perhaps readers sixty years from now will find today's model letters quaint. Like archaeologists, technical and business writing specialists may find through this bibliography relics of the past that will give them insight about the future.

Today, much attention is being given to the issue of sexism--especially in business letters. No business writing book today is likely to advocate a principle such as the one John McCloskey propounds in Handbook of Business Correspondence (1932): "... base letters to men on logic and conviction, " but for women "use persuasion and emotional appeals rather than logical argument. " Yet in 1971, Robert M. Archer and Ruth Pearson Ames in Basic Business Communication warn the reader in their discussion of business-communication psychology that women "will say yes when they mean no or no when they mean maybe. " In a strikingly modern flavor, on the other hand, Eleanora Banks in Putnam's Correspondence Handbook (1919) mentions omitting the salutation altogether as an option in solving the problem of addressing letters to women, but the author laments, "to make the matter more complicated, authorities do not wholly agree" on the solution.

In addition to examining works to gain historical perspective, scholars might examine them to find subjects for future study. Patrick Meredith analyzes the process of scientific writing from the prespectives of epistemology, psychology, linguistics, semantics, psychophysics, and pragmatics in Instruments of Communication: An Essay on Scientific Writing (1966). The American Business Communication Association's Guidelines for Research (1977) also provides future directions for research. Other works suggest further directions for study in such widely diverse areas as the use of playscript procedure (Matthies, 1961), the study of controversial subjects in company publications (Dover, 1959), the attitudes that produce poor business writing (Whyte et al. , 1952), and the development of behavioral objectives for courses in vocational writing (Barton, 1970). Textbook writers might profitably examine the pedagogical apparatus of past, as well as current, textbooks. Some early texts, of course, are not as sophisticated as many of today's, but they could provide examples from which future authors could build.

Cultural Importance

In the anthology English and Engineering (1917), Frank Aydelotte aims to help engineering students understand that their work "is bound up with the spiritual advancement of the race--with the world of science, of literature, and of moral ideals." Clyde W. Park in English Applied to Technical Writing (1926) stresses the development of the technical writer's literary background and rejects the idea of a narrow view for the technical writer. J. Harold Janis in Business Communication Reader (1958) includes sections on "Moral Values in Business" and "The Businessman in American Literature." Josephine Baker asserts in Correct Business Letter Writing and Business English (1911) that a "business man's letter betokens either his illiteracy or his culture." Babenroth and Viets in Readings in Modern Business Literature (1928) say that they have collected essays that stress the importance of a liberal-arts background for the business student. To help students develop a sensitivity to language, Louise Roberts in How to Write for Business (1978) asks them to write a modern version of "A Message to Garcia," analyzing modern attitudes toward workmanship. Many other authors have suggested that business and technical writing must not become unthinkingly mechanical or behavioristic.

Numerous authors have pointed to the literacy of technical writers of the past--people, like Charles Darwin and Thomas H. Huxley. Others (e.g., Bower and Mazzeo, 1979) have shown that scientists, such as Loren Eiseley and Lewis Thomas, can create exciting prose. Perhaps none has gone as far as George E. Williams (1948), who argues that technical writing can be considered a literary form: "If the language is clear and convincing, if all the parts of the subject-matter are interrelated to form an organic whole which is something more than the mere sum of the parts, the product may rank as literature; if, in addition, the information is accurate and the argument logical, it may properly be described as 'technical literature.'"

Of course, few authors represented in this bibliography would suggest that the products of business and technical writing will survive as cultural landmarks. Most would agree, however, that they are important to the extent that they enlighten us on the complex and crucial issues of the day and help us conduct our business with some humanity toward our fellow human beings.

B1 Annotated Bibliography on Technical Writing, Editing, Graphics
     and Publishing 1950-65. Hollywood, Calif. : Western Peri-
     odicals, 1969.
This bibliography is no longer in print and is unavailable. It was
compiled by California members of the Society for Technical Communication and circulated regionally. Many of the items listed were taken
from an earlier published bibliography edited by Theresa Ammannito
Philler and others.

B2 Balanchandran, Sarojini. Employee Communication 1965-1975.
     Urbana, Ill. : American Business Communication Associa-
     tion, 1975. 55p.
Some 675 items--mostly articles and a few books and dissertations--
deal with employee communications. The author aims to provide a
listing of materials that treat the communication gap between management and employees. Sources for the bibliography are Business
Periodicals Index, Public Affairs Information Service Bulletin, Accountant's Index, Personnel Literature, Personnel Management Abstracts, Dissertation Abstracts, Library of Congress Subject Catalog,
Applied Science and Technology Index, Readers' Guide to Periodical
Literature, Psychology Abstracts, Sociological Abstracts, and Work
Related Abstracts. The subjects covered are communication in management, communication in personnel management, reports to employees, attitude surveys, employee publications, bulletin boards,
employee evaluation and ratings, and employee motivation and training. About 20 percent of the works cited are annotated with one sentence or phrase. Entries are in a bibliographic form used in the
sciences. There are no divisions by subject.

B3 Balachandran, Sarojini. Technical Writing: A Bibliography.
     Urbana Ill. , and Washington, D. C. : American Business
     Communication Association and Society for Technical Com-
     munication, 1977. 147p.
This bibliography cites mostly articles and papers, with some books
on technical writing published since 1965. Sources for the bibliography are Applied Science and Technology Index, Biological and

Agricultural Index, Books in Print, British Technology Index, Current Contents, Current Index to Journals in Education, Education Index, Engineering Index, ERIC Index, and the Library of Congress
Subject Catalog. The subject titles covered are communication of
technical information, engineering writing, report writing, science
writing, technical manuals, technical reports, and technical writing.
Most of the works cited have brief annotations giving the content
covered. The bibliography is arranged alphabetically by author, and
entries are in the bibliographic form used in the sciences. There
are no divisions by subject; however, a subject index is provided.

B4   Ball, Sarah B.   1600 Business Books. 2nd ed. revised and
              enlarged to 2,100 titles by L. H. Morley and S. H. Powell.
              New York: H. W. Wilson, 1917. 232p.
This bibliography's first edition was compiled by Ball, who began
the second enlarged edition, which was completed by Morley and
Powell. Based on the holdings of the Business Branch of the Newark, New Jersey, Public Library, the bibliography was intended "to
give the business men of Newark a comprehensive yet simple guide
for office use. " Books are cited by author, title, and by subject,
all in one alphabetical listing. Books on writing are grouped primarily under "Commercial Correspondence" and special topics, such as
"House Organs. " There are no annotations. A list of publishers
with their addresses is included.

B5   Bibliography of Publications Designed to Raise the Standard of
              Scientific Literature. Paris: United Nations Educational,
              Scientific, and Cultural Organization, 1963. 83p.
The purpose of this bibliography, which lists 354 works in several
languages including English, Russian, German, and French, is to
help authors prepare manuscripts; compose, draft, and edit scientific
and technical texts; collect materials and references; and compile indexes. A majority of the works cited have brief annotations describing their subject matter. The bibliography has six divisions. Three
of these are particularly related to technical writing: "General Works
on Language and Composition, " "The Technique of Technical Writing, "
and "Readings in Science for Technical Authors. " The bibliography
concludes with a general index and an index listing the works by language.

B6   Bowman, Mary Ann. "Books on Business and Technical Writing
              in the University of Illinois Library. " The Journal of Busi-
              ness Communication 12 (Winter, 1975), 33-67.
This bibliography lists, alphabetically by author, books, monographs,
and dissertations that appeared between 1950 and 1973. None of the
entries is annotated, and there are no divisions by subject. The
titles of books are not differentiated from the rest of the bibliographical matter either by italics or other typeface.

B7   Bowman, Mary Ann, and Stamas, Joan D. Written Communica-
              tion in Business: A Selective Bibliography, 1967-1977.
              Champaign, Ill. : American Business Communication Asso-
              ciation, 1980. Index. 101p.

This bibliography lists over 800 articles and books dealing with such topics as business letters, memos, reports, résumés, employee publications, and direct mail. Topics dealing with technical writing or word processing are not included. Most of the entries are for journal articles. The annotations consist of a sentence or phrase giving the focus of the work. A subject index is included. Coauthors are cited in the main listing with a cross-reference to the entry for their work.

B8   Burkett, Eva M. , comp.   Writing in Subject-Matter Fields: A
         Bibliographic Guide, with Annotations and Writing Assign-
         ments.   Metuchen, N. J. :   Scarecrow, 1977.   201p.
This bibliography is divided into seven sections: "About Writing";
"Writing and Literature"; "Writing in History, Autobiography and
Biography, Law"; "Writing in Science"; "Technical and Business Writing"; "Interdisciplinary Writing"; and "Writing Articles for Newspapers and Magazines: Travel, People, Places, Occasions, etc. "
Each section has a one-paragraph introduction explaining the importance of clear writing for that particular field.   Entries cover books
and magazine articles and are annotated with a brief paragraph.
Each section ends with writing assignments for students.   The author
suggests that the bibliography be used in courses taught by experienced teachers.   An author-title index is included.

B9   Carter, Robert M.   Communication in Organization: An Anno-
         tated Bibliography and Sourcebook.   Management Informa-
         tion Guide 25.   Detroit: Gale, 1972.   272p.
This bibliography, one of a series of management information guides
to resources in various fields, is aimed at three categories of readers--practicing managers, business people, or engineers; students of
communication; and librarians and information specialists who serve
these groups.   The over 1, 200 items cited include books, periodicals,
parts of books, some films, and media.   The author includes an annotated bibliography, a listing of addresses of periodicals and book
publishers, an author index, a title index, and subject index.   The
one-or-two-sentence annotations explain the main subject of each
work.   Occasionally, a specific reader is identified.   However, no
information on the structure, approach, or usefulness of the work is
provided.   The bibliography section is divided into nine major subject divisions: "Theories and Systems of Organizational Communication, " "Barriers to Organizational Communication, " "Vertical Communication, " "Horizontal Communication, " "Communication Media, "
"Information Communication Channels, " "Organizational Change, "
"Evaluation of Effectiveness of Organizational Communication, " and
"Sourcebooks and Articles. "   Within each major division, subsections further classify the works by minor subjects and list them
chronologically.

B10   Cunningham, Donald H. , and Hertz, Vivienne.   "An Annotated
         Bibliography on the Teaching of Technical Writing. "   Col-
         lege Composition and Communication 21 (May, 1970), 177-
         186.
This bibliography, which annotates 67 journal articles published be-

tween 1954 and 1969, is in three parts: "Definitions and Distinctions," "Course Descriptions and Specific Assignments," and "Resource Materials." The periodical sources used for this bibliography include journals in science and technology (and their pedagogy). Each annotation gives the basic content and organization of the article and evaluates its usefulness for the instructor. The articles selected provide teachers of technical writing with suggestions for teaching, descriptions of successful technical writing courses, and locations of resource materials.

B11 De Vergie, Adrienne. English for Lawyers: A Bibliography
         of Style Manuals and Other Writing Guides. Austin: University of Texas at Austin, 1975. 47p.
This unannotated bibliography lists books on writing, grammar, and composition to help the beginning lawyer or law student write clear prose. The first three pages cite books that have a specific reference to legal writing. The main bibliography lists handbooks of style, grammar, and rhetoric. Most of the books included are college texts.

B12 Donovan, Robert B., ed. Technical Writing Texts for Second-
         ary Schools, Two-Year Colleges, and Four-Year Colleges.
         Urbana, Ill.: National Council of Teachers of English,
         Committee on Technical and Scientific Writing, April
         1978. 12p.
This bibliography lists and annotates 59 texts and anthologies. Only titles currently in Books in Print were included. Annotations range from one sentence to one paragraph. The entries are coded for the appropriate level: "(1)" indicates usefulness in secondary schools; "(2)" indicates potential use in two-year programs; "(3)" indicates use in junior-senior levels of four year colleges or beyond. Prices of the books are included.

B13 Estrin, Herman A. "Bibliography" in The Teaching of College
         English to the Scientific and Technical Student. Mimeo-
         graphed. Urbana, Ill.: National Council of Teachers of
         English, 1963. 8p.
Available only on microfilm as an Educational Resources Information Center (ERIC) document, #ED 020 197, this bibliography is part of a NCTE report by the Committee on College English for the Scientific and Technical Student. There are three divisions: "Suggestions for More Effective Writing," "Technical Writing Programs," and "Technical Writing in the Professional World." The first division is further divided into "Articles," "Books," "General Dictionaries," "General Style Manuals," "Technical Aids," and "Graphic Aids." There are no annotations.

B14 Falcione, Raymond L., and Greenbaum, Howard H. Organiza-
         tional Communication 1976: Abstracts, Analysis, and
         Overview. Urbana, Ill., and Austin, Tex.: American
         Business Communication Association and International
         Communication Association, 1977. 300p.
The third annual edition of this annotated bibliography contains 868

abstracts of works in organizational communication published from
October 1, 1975, through September 30, 1976. The scope has been
expanded to include doctoral dissertations on industrial and social
psychology. This edition includes a 78-page review of the literature
and an appendix that explains the research methods used by the editors. An author index is also provided. Two new indexes not included in the earlier editions cover the abstracts of field studies.
One index indicates the type of organization from which the researcher obtained data, and the other index indicates the sources used by
the researcher. The divisions that include material on writing are
"Skill Improvement and Training in Organizational Communication";
"Communication Media in Organizations: Software and Hardware";
and "Texts, Anthologies, Reviews, and General Bibliographies Relative to Organizational Communication. " Each division is further divided into (1) books and dissertations, and (2) articles, papers, and
U. S. government publications. The coding for research characteristics of each entry, which had been dropped from the second edition,
has been restored, and cross-referencing has been provided.

B15   Falcione, Raymond L. , and Greenbaum, Howard H. Organization-
       al Communication Abstracts--1975. Urbana, Ill. , and Aus-
       tin, Tex. : American Business Communication Association
       and International Communication Association, 1976. 142p.
This second annual edition contains more than 700 abstracts of works
published from October 1, 1974, through September 30, 1975. The
two-or-three-sentence annotations cover the main subject of each
work. The divisions that include material on writing are "Skill Improvement and Training in Organizational Communication"; "Communication Media in Organizations: Software and Hardware"; and "Texts,
Anthologies, Reviews, and General Bibliographies Relative to Organizational Communication. " Each section is further divided into (1)
books and dissertations, and (2) articles, papers, and U. S. government publications. The coding used in the first volume as to whether
the work is laboratory study, field study, or nonempirical work has
been dropped. This bibliography contains an author index for the
1975 and 1976 volumes. The references to dissertations include the
volume and page numbers from Dissertation Abstracts International.

B16   Goldberg, Jay J. "A Survey of Scholarly Works in Technical
       Writing. " Technical Communication 22 (First Quarter,
       1975), 5-8.
This article reviews scholarly literature in technical writing from
1965 to 1973. The author discusses only 35 items because he feels
there is a "dearth of scholarly works" in technical communication.
The fields surveyed are (1) readability, (2) profiles (professional
training and job categories and activities), (3) information transfer,
and (4) miscellaneous professionally oriented studies.

B17   Greenbaum, Howard H. , and Falcione, Raymond L. Organiza-
       tional Communication Abstracts--1974. Urbana, Ill. , and
       Austin, Tex. : American Business Communication Asso-
       ciation and International Communication Association, 1975.
       87 pages.

This first edition contains more than 400 entries published from
October 1, 1973, through September 30, 1974. The bibliography includes a review of the research methods used by the compilers and
a statement of the scope of coverage. The brief two-or-three sentence
annotations give the main subject of the book or article. The entries
are grouped into nine divisions. The three divisions that include material on writing are "Skill Improvement and Training in Organizational Communication"; "Communication Media in Organizations: Software and Hardware"; and "Texts, Anthologies, Reviews, and General
Bibliographies Relative to Organizational Communication." Topics
included in each classification are listed at the beginning of each section. Each section is further divided into (1) books and dissertations,
and (2) articles, papers, and U.S. government publications. Entries
are coded as laboratory studies, field studies, or nonempirical
works.

B18   Greenbaum, Howard H., and Falcione, Raymond L.   Organiza-
        tional Communication 1977: Abstracts, Analysis, and
        Overview.   Champaign, Ill., and Austin, Tex.: Ameri-
        can Business Communication Association and International
        Communication Association, 1979.   297p.
This fourth edition includes 469 books and 478 articles and papers.
An 82-page overview of the literature covers works from October 1,
1976, to September 30, 1977. There are nine classifications, each
divided into (1) books and dissertations, and (2) articles, papers,
and U.S. government publications. Entries are coded for research
characteristics (field study, prescriptive/descriptive writing, theoret-
ical/conceptual writing). Cross-referencing is provided along with
an author index, index of field studies and type of organization, and
an index of data-collection methods. The sections including material
on writing are "Skill Improvement and Training"; "Communication
Media: Software and Hardware"; and "Texts, Anthologies, Reviews,
and General Bibliographies." New to this edition is the inclusion of
Ph.D. dissertations in the field of industrial and social psychology.

B19   Hertz, Vivienne, and Cunningham, Donald H.   "Bibliography of
        Police Report Writing," Police Chief 38 (August, 1971),
        44, 49-50.
This bibliography, consisting of 102 items, is "a modest attempt at
describing the state of the art of police report writing." Listed are
nine books on police-report writing, along with books containing some
material on the subject, bound and unbound manuals and bulletins,
and articles. None of the citations is annotated. Several categories
of works are omitted: unavailable manuals produced by local police
departments, teaching outlines and syllabi, and articles from foreign
journals.

B20   Lytel, A. H.   Bibliography on Technical Writing and Related
        Subjects.   Syracuse, N.Y.: General Electric Company,
        Electronics Laboratories, 1958.

B21   McClure, L.   "Two Bibliographies: Technical Writing Books
        in Print; Phototypesetting." IEEE Transactions on En-

gineering Writing and Speech 8 (December, 1965), 67-70.
The first of these two bibliographies is an unannotated list of 58 books
on technical writing. The works cited emphasize scientific- and en-
gineering-report writing, proposals, specifications, and style sheets;
the bibliography is therefore useful primarily for technical people
(especially engineers and chemists). The second of these two bibliographies covers only phototypesetting.

B22 Morley, Linda H., and Kight, Adelaide C. 2, 400 Business
     Books and Guide to Business Literature. New York: H.
     W. Wilson, 1920. 456p.
This bibliography is an enlarged version of the first and second editions of 1, 600 Business Books. The contents are based on the holdings of Business Branch of the Newark, New Jersey, Public Library.
Books are cited under the author, title, and by subject, all in one
alphabetical listing. Books on writing are grouped primarily under
"Commercial Correspondence" and "Report Writing, " although other
categories may contain pertinent citations. Major categories are
subdivided. There are no annotations. A list of publishers and
their addresses is included.

B23 Philler, Theresa Ammannito; Hersch, Ruth K.; and Carlson,
     Helen V.; eds. An Annotated Bibliography on Technical
     Writing, Editing, Graphics, and Publishing 1950-1965.
     Washington, D.C., and Pittsburgh: Society of Technical
     Writers and Publishers and Carnegie Library of Pitts-
     burgh, 1966. 312p.
This bibliography contains annotations of 500 books and 1, 500 articles
published from January 1, 1950, to December 31, 1965. The editors'
scope is broad, since "the needs of technical writers, editors, illustrators, and publishers vary." The annotations include material from
local-chapter bibliographies of the Society of Technical Writers and
Publishers (now the Society for Technical Communication). The work
has three main divisions: the bibliography, a permuted title index
that serves as a keyword subject index, and an author index. The
bibliography lists the books and articles separately. The brief,
numbered annotations identify the primary audience and describe the
essential content of each work cited, but they provide little information on approach or structural design. A final section lists titles
and addresses of periodicals cited in the annotations. The book is
computer printed in all-capital letters with no punctuation other than
periods and commas.

B24 Ronco, P. G., et al. Characteristics of Technical Reports
     that Affect Reader Behavior--A Review of the Literature.
     Springfield, Va.: Clearinghouse for Federal and Techni-
     cal Information, 1966. 191p.
This annotated bibliography covers 411 books, articles, and papers.
It is part of a report surveying the characteristics of technical reports that affect reader behavior. Most of the items cited, therefore, are concerned with readability studies. The paragraph-length
annotations give the objectives of the study, the test procedures, and
the results. For works that do not deal with specific experiments,

the annotations give the general subjects and main points covered.
There are no divisions by subject.

B25   Van Veen, F.   "An Index to 500 Papers through 1962 on En-
            gineering Writing and Related Subjects. " IEEE Trans-
            actions on Engineering Writing and Speech EWS-6 (Sep-
            tember, 1963), 50-58.
This bibliography lists over 500 papers, some unpublished, that have
been delivered at professional meetings.   The bibliography covers
material on writing, on speaking, and on publishing technical information.   The author states that "this index is not necessarily 100%
exhaustive; some titles may have been omitted, and some misclassi-
fied. "   No annotations of the papers are provided.   The bibliography
is divided into nine sections: "The Technical Writer and His Profession, " "Technical Publications, " "Graphic Arts, " "Corporate Aspects, " "Economic and Political Aspects, " "Communication Process, "
"Writing and Editing, " "Oral Presentation, " and "Documentation. "
These main sections contain numbered subsections that further classify the papers by subject.   An author index is included.

B26   Walsh, Ruth M. , and Birkin, Stanley J. , eds.   Business Com-
            munication: An Annotated Bibliography.   Westport, Conn. :
            Greenwood, 1980.   686p.
This bibliography lists 1, 657 "journal articles, books, and dissertations written in English during the 1960s and 1970s by researchers
and practitioners on the subject of business communication. " "Business communication" is interpreted broadly, since many of the items
listed include works on communication, writing, mass media, and
linguistics in general.   The book is divided into three parts.   Part
I (132 pages), "Author/Title Listing, " lists alphabetically all authors
and coauthors with cross-references to the titles of their works.
Part II (375 pages), "Key-Word-in-Context Listing, " indexes all entries by keywords in their titles.   Part III (279 pages), "Abstracts, "
lists works alphabetically by author.   Without explanation, however,
the section is divided into two alphabetical groups, one ending with
Zinsser (entry C0906) on page 561 and the second beginning immedi-
ately with American Business Communication Association (C0907)
and Adler (C0908) on the same page.   The entries contain publication data and in some cases annotations, which range from a single-
sentence description to excerpts from book reviews.   The book is
computer printed, thus the punctuation is limited and each word appears in all capital letters.

B27   Walsh, Ruth M. , et al. , eds.   "Business Communications: A
            Selected, Annotated Bibliography. "   The Journal of Busi-
            ness Communication 11 (Fall, 1973), 65-112.
This bibliography, containing 454 books and articles published between 1940 and 1972, lists selected works on business communication
and allied areas, such as mass media, public relations, research
methodology, technical reporting, rhetoric, and semantics.   The emphasis is on report writing and business correspondence.   The articles are selected from journals and government publications on business communication, enginneering, and management.   Many works

have at least a one-sentence annotation identifying the basic content. The longer annotations cover subject and organization and the work's value for the intended reader.   The entries are arranged alphabetically by author.

1 Adamson, Donald, and Bates, Martin. Nuclear English for Science and Technology: Biology. London: Longman, 1977.

2 Adamson, Vera, and Lowe, M. J. B. English Studies Series, General Engineering Texts. London: Oxford University Press, 1973. Index. 203p.
This book is a collection of 20 essays on "the kind of material which first year undergraduates and research workers from overseas may read or listen to in the lecture rooms, libraries and laboratories of a typical university in the United Kingdom." The essays come from such sources as textbooks, lectures, and scholarly papers and are on such topics as nuclear power, automation, aircraft structures, and catalytic cracking. Each essay is followed by several questions and exercises that test comprehension and grammatical proficiency. For example, the reader may be asked to "complete the following sentences with the present participle or past participle form of the verbs in parentheses."

3 Addington-Symonds, F. Commercial Correspondence. New York: McKay, 1973.

4 Adelstein, Michael E. Contemporary Business Writing. New York: Random House, 1971. Index. 365p.
This textbook is designed for college students at the sophomore level or above who have taken a basic writing course. Since the author stresses the process of writing, he does not attempt to cover the theory of business communication. He does, however, discuss in the first chapter the importance of business letters and reports. Thorough chapter-length treatment is provided for rhetorical topics, such as "words," "usage," "brevity," "vigor," "clarity," and "style." These chapters include illustrations and many exercises. Only the final chapter is devoted to forms of letters, reports, and instructions. The end-of-chapter summaries provide an overview of the topics covered.

5 Agg, Thomas R., and Foster, Walter L. The Preparation of

Engineering Reports.   New York: McGraw-Hill, 1935.   Index.   Bib.   192p.

This textbook is designed for use in short courses for engineering
students as a supplement to standard English courses.   The authors
include an outline for a course in engineering report writing in Appendix C.   The book covers such topics as the characteristics of
the engineering report, formal and informal reports, organization,
collection of data, style and arrangement, form report, and visual
aids.   Appendix A is a handbook on punctuation and capitalization.
Appendix D is a bibliography of reference books, and Appendix B
lists abbreviations commonly used for technical terms.   Excerpts
from company reports are included.   There are no exercises.

6   All About Letters.   Washington, D. C.:   U. S. Postal Service,
      1979.   64p.

This guide for high school students was produced in cooperation with
NCTE.   The 20 sections include discussions of the importance of letter writing, consumer letters, college-application letters, job-
application letters, letter format, letters of inquiry, and postal information.   Included is a 25-page section on where to find information on a variety of items, such as career exploration, survival kits,
national parks, and weather information.

7   Allbutt, Sir. T. Clifford.   Notes on the Composition of Scientific
      Papers.   3rd ed.   1923; rpt.   London: Macmillan, 1925.   In-
      dex.   192p.

This book, originally published in 1904, was intended as a guide for
medical students preparing theses.   Chapter 1, "Introductory, " discusses finding the thesis subject, references, openings and endings,
and word choice.   Chapter 2, "On Composition, " covers grammar,
vocabulary, abstract style, punctuation, and emphasis.   It includes
a list of words for which the author explains desired usage.   For
example, "female" is described as "disagreeably zoological. "

8   Allen, J. P. B. , and Widdowson, H. G.   English in Physical
      Science.   London: Oxford University Press, n. d.

9   Althouse, Calvin Osborne.   Business Letters.   1910; rpt. Phila-
      delphia: Penn, 1924.   208p.

According to the author, "one of the great needs of the business
world today is to find those who can indite a good business letter,
cordial in tone, correctly punctuated and expressed in clear unmis-
takable English. "   This book has 18 chapters covering such topics
as effective style, parts of a letter, format and folding letters, types
of letters, postal information, and various other business forms,
such as bank drafts and billing notices.

10   Altmaier, Carl Lewis.   Commercial Correspondence and Postal
       Information.   rev. ed.   New York: Macmillan, 1913.   In-
       dex.   252p.

The first edition of this textbook was published in 1904.   The 16
chapters cover such topics as the need for a course in commercial
correspondence; format; and the need for clarity, completeness,

courtesy, and brevity in correspondence.   Nine chapters are devoted
to types of letters: inquiry, orders, requests for payment, introduction, recommendation, sales, telegrams, contracts, and personal
letters.   One chapter covers filing and indexing, and one chapter
contains a variety of sample letters.   Chapter 16, "Postal Information," is a lengthy discussion of the history of the post office, as
well as the organization and services of the office.   There are writing assignments and cases for study.

11   Alvarez, Joseph A.   The Elements of Technical Writing.   New
        York: Harcourt Brace Jovanovich, 1980.   Index.   208p.
Although this book contains writing assignments at the end of its chapters, it could be considered a handbook as well as a textbook since
sections within the eight chapters are numbered for easy reference.
The chapters cover general principles of technical writing, word
choice, sentence construction, organization, punctuation, format,
style, and the writing process.   There is a detailed section on research sources.   Four appendixes follow the chapters and include the
analysis of a sample report, reviews of grammar and spelling, and
metric-conversion tables.

12   Ammon-Wexler, Jill, and Carmel, Catherine.   How to Create
        a Winning Proposal.   Santa Cruz, Calif. :   Mercury Commun-
        ications, 1976.   238p.
This book, which is a looseleaf binder, is written for consultants,
business executives, and development directors of nonprofit organiza-
tions--all those who write proposals.   The five sections are "Man-
aging the Proposal Effort, " "The Basic Components of a Proposal, "
"Writing the Proposal, " "Specialized Proposals, " and "Government
Proposals. "   The 19 chapters cover such points as "Planning and Organizing, " "The Management Component, " "The Pricing Component, "
"The Winning Technical Proposal, " "The Winning Management Proposal, " "Government Contracting, " and "The Winning Grant Proposal. "
The appendix is a comprehensive list of information sources, such as
"Atomic Energy Communication Sources" and "NASA Information
Sources. "   Writing style in the book is informal.   Each chapter begins with a colored divider sheet that lists the items covered in the
chapter.   Each chapter is paged separately.   At the end of the chapter is a checklist of "Points to Remember. "

13   Anderson, C. R. ; Saunders, Alta Gwin; and Weeks, F. W.
        Business Reports: Investigation and Presentation.   3rd ed.
        New York: McGraw-Hill, 1957.   Index.   407p.
The 16 chapters in this textbook cover such topics as report language, figures, and collecting data.   Chapters 13-16 are concerned
with writing the individual elements of the report from cover to index.   The chapters focus on the steps in the report writing process
in logical order, giving thorough attention to form and content.
Chapters 4 and 5 cover credit, letter, and memo reports.   Appendix
I is an example of a student report.   Appendix II is a set of 43 writing assignments for the student.   The assignments cover various
types of reports.

14 Anderson, W. Steve, and Cox, Don Richard. The Technical
        Reader: Readings in Technical, Business, and Scientific
        Communication. New York: Holt, Rinehart and Winston,
        1980. Index. 378p.
This anthology contains over 70 selections divided into three main
parts. Part I includes articles about report writing and is divided
into three sections. Section 1 includes general discussions, such
as "The Modern Report," by Jacques Barzun and Henry F. Graff,
and "Credo of a Tech Writer," by John Frye. Section 2 includes
articles on "Language, Style, and Audience," such as "Audience
Analysis," by Thomas Pearsall, and "The Capacity to Generate Language Viability Destruction," by Edwin Newman. Section 3 treats
problem solving and includes "The Effective Decision," by Peter
Drucker. Part II contains articles that illustrate the use of description, narration, illustration/example, process, comparison/contrast,
definition, classification/division, cause-effect, and analogy. Part
III provides "specific examples of reporting" related by methods of
development, persuasion and instructions. Following each reading
are study questions.

15 Andrews, Clarence A. Technical and Business Writing. Boston:
        Houghton Mifflin, 1975. Index. 243p.
Although letters are included, this textbook is primarily geared toward a one-semester report-writing course. The 14 chapters move
from simple telephone and letter reports to writing for professional
journals. The topics covered include technical sentence pattern,
letters, technical description, informal and formal reports, writing
for magazines, and retrieving technical information. There are study
questions after each chapter and numerous outside writing assignments. The book includes samples of reports, technical illustrations,
and graphs. Six appendixes offer samples of "Good Student-Written
Instructions," "The Formal Report," "The Progress Report," "A
Student Feasibility Report," "The Proposal Report," and "The Report
Based on Literature Research."

16 Andrews, Deborah C., and Blickle, Margaret D. Technical
        Writing: Principles and Forms. New York: Mac-
        millan, 1978. Index. Bib. 431p.
This book reflects the authors' rhetorical approach to the subject of
technical writing. It starts with an introduction covering the principles of reader and objective and the writing process. Section I,
"Evidence," covers research methods and sources. Section II, "Expression," covers technical style, organization, methods of discourse,
and graphics. Section III, "Forms," covers abstracts, proposals,
progress reports, articles, and correspondence, in addition to format
conventions. Section IV, "Mechanics," serves as a handbook. Appendix A is "How to Document an Article or a Report," and Appendix B is "Selected Sources on Scientific and Technical Writing."

17 Archer, Robert M., and Ames, Ruth Pearson. Basic Business
        Communication. Englewood Cliffs, N.J.: Prentice-Hall,
        1971. Index. 401p.
The 12 chapters of this textbook are divided into three parts. Part

I, "Principles of Communications," covers the psychology of business, courtesy in business communication, readability and style,
word choice, grammar, and organization.   Part II, "Communications
in Action," covers types of business communications: sales, inquiries, orders, complaints, credit and collections, personal messages
in business (telephone techniques, letters of condolence, etc.), and
applications.   Part III, "Intra-Company Communications," covers internal memos and reports.   Exercises and problems appear at the
end of most chapters.   The two appendixes cover "frequently confused
words" and "words and phrases with undesirable overtones" and alternatives.   The book assumes that the reader is already in a business situation and is a male.

18 Arnold, Christian K.   Technical Writing Manual.   Cleveland:
        Electronic Periodicals, 1959.   Unp.
This book is a glossary of about 60 pages covering correct usage.
The book resulted from a series of "Technical Writing Seminars"
published in Electronic Digest.   Requests for copies of the monthly
compilations of correct usage resulted in the decision to publish the
lists in a monograph.   There are about 250 usage entries, from
"ability capacity" to "while."

19 Arny, Mary Travis, and Reaske, Christopher R.   Ecology: A
        Writer's Handbook.   New York: Random House, 1972.
        Bib.   112p.
This book is geared to the ecology student who is concerned about
writing and to the layperson who is writing letters to the editor about
ecology.   It is not a text for a writing course but could function as
a supplement in an ecology course.   The seven chapters cover such
topics as the importance of writing about ecology, selecting the topic,
restricting the topic, technique, style, and persuasion.   Chapter 5
is a 25-page "Glossary of Ecological Terms."   Chapter 6 is a usage
guide, and Chapter 7 covers grammar and mechanics.   Illustrated
with photos of wildlife, the book includes sample letters to corporations and newspaper editors.

20 Aspley, J. C., ed.   Salesman's Correspondence Manual.   2nd
        ed.   Chicago: Dartnell, 1917.   97p.
This reference manual for salespeople interested in writing better
letters is based on the "experience and views of no less than one
hundred and eighty-five sales executives subscribing to the Dartnell
Sales Research Service."   The 12 chapters are divided into three
parts.   Part I, "Letters to the Office," includes general correspondence, reports to the office, letters of complaint, credit letters, and
advertising letters.   Part II, "Letters to Your Customers," is devoted to general communications, advance letters, and handling complaints.   Part III, "Circular Sales Letters," covers general sales
letters, letters for customer's use, collection letters, and a section
on "don't"s for letter writers.   The book includes a section on "Words
Commonly Misused."

21 Atkinson, Philip S., and Reynolds, Helen.   Business Writing and
        Procedures.   New York: American Book, 1970.

22 Atthreya, N. H. Modern Correspondence: How to Increase Ef-
         ficiency and Reduce Costs. Bombay, India: MMC School
         of Management, 1972. Bib. 239p.
This book is unconventional in that the sentences are set like the
lines of a poem. There is much white space and the lines start at
odd places on the pages. Although there are 239 pages, the text is
relatively slight. The 63 chapters are divided into 13 parts. The
aim of the book is to build the desired attitude and enthusiasm toward commercial correspondence that will develop effectiveness.
The objectives of letters are discussed along with tested approaches.
Much attention is paid to cost and how to reduce it. The fundamentals of writing and special types of letters are also covered.

23 Ault, Nelson A. , and Magill, Lewis M. Business Letters. Los
         Altos, Calif.: Chandler, 1956. 35p.
This work has eight brief chapters divided into several sections.
Each section opens with advice followed by an illustration of the
principle. For example, one section in Chapter 1 advises, "Avoid
Jargon. " This advice is followed by an example of jargon. Topics
covered by chapters include effective writing style, parts and forms
of the business letter, adapting the correspondence to the reader,
sales letters, applications letters, claim and adjustment letters, and
credit and collection letters. Chapter 8, "Miscellaneous Letters, "
contains more examples.

24 Aurner, Robert R. , and Burtness, Paul S. Effective English
         for Business Communication. 6th ed. Cincinnati: South-
         Western, 1970. Index. 680p.
This text, primarily geared to the high school student, consists of
69 chapters divided into 16 units in three divisions. The five goals
for the students using the book are to develop accuracy and sensitivity to words, learn organization, become familiar with business procedures and principles, increase understanding of business problems,
and deepen understanding of people's interests, motives, and prefer-
ences. Division I, "The Resources of English, " distinguishes between written and spoken English and presents the basic principles
of communication and reviews the basic sentence patterns. Division
II, "Effective Business Messages, " covers such topics as tone, organization, and style. Division III, "The Reference Guide, " covers
questions of grammar and punctuation, including a section on sentence diagramming. There are student exercises, writing assignments, and case studies.

25 Austen, David, and Crosfield, Tim. English for Nurses. Lon-
         don: Longman, 1976. 138p.
This textbook is a very basic practice manual for nursing students.
The 24 units cover word choice on such topics as "Parts of the
Body, " "Sterile Procedures, " "Admissions, " "Urine, " "Pulse. "
The unit contains definitions for words and fill-in exercises using
those words. Verb choice is explained and sentences requiring word-
choice variety are included. Exercises provide practice verb-tense,
in polite requests, and in antonyms. The "Word List" at the end is
a glossary of technical terms. Since there is no preface, it is not
completely clear what the overall purpose of this textbook is.

26 Automatic Letter Writer. Chicago: Shaw, 1914.

27 Avery, James K. , et al. , eds.   Report-Writing in Dentistry--
        A Teaching Outline. 9th ed. Ann Arbor, Mich. : Over-
        beck, 1960. Bib. 70p.
Aimed at helping students in dentistry prepare research papers, this
style guide is divided into seven parts: "The Use of the Dental
Library--A Specialized Skill, " "Scientific Reporting, " "The Evaluation of Information, " "Reporting References in Dental Writing, " "Abbreviations, " "Style and Construction in Scientific Writing, " and
"Bibliography. " Topics include using the card catalog, typing the
thesis, writing an abstract, documentation, and avoiding verbosity.
The guide is edited by four dentistry teachers, an English teacher,
a librarian, two statisticians, and a stenographer, all of whom are
identified (except the stenographer).

28 Avett, Elizabeth M. Today's Business Letter Writing. Engle-
        wood Cliffs, N. J. : Prentice-Hall, 1977.

29 Aydelotte, Frank, ed. English and Engineering. 2nd ed. New
        York: McGraw-Hill, 1923. Index. Bib. 415p.
Originally published in 1917, this anthology of 28 articles purposed
"to teach the student to write not by telling him how, not by doing
his thinking for him, but by stimulating him to think for himself
about his own problems, about his work and its place in the world. "
The collection also aimed at leading the student to think about engineering as "one of the liberal professions" and to see "how his
work of designing material conveniences for men is bound up with
the spiritual advancement of the race--with the world of science, of
literature, and of moral ideals. " Accordingly, the articles are divided into five sections: "Writing and Thinking, " "The Engineering
Profession, " "Aims of Engineering Education, " "Science and Literature, " "Literature and Life. "

30 Babenroth, A. Charles. Modern Business English. 5th ed. Revised
        by Charles Chandler Parkhurst. New York: Prentice-
        Hall, 1955. Index. 645p.
The 14 chapters of this textbook are divided into three parts. Part
I, "Fundamentals of Effective Communication, " covers the telephone
and telegram, principles of business writing, effective style, and
planning and layout of letters. Part II, "Types of Business Communication, " deals with types of letters. There is one chapter on reports, and there are three chapters on sales letters. Part III, "Reference Section, " is a grammar, punctuation, and usage handbook
with sentence exercises. Included is a list of forms of address and
tips on using the telephone and telegram. Part IV, "Questions and
Problems, " is a 124-page section giving discussion questions and
writing problems keyed to the chapters. There is a lengthy list of
companies who contributed to the book.

31 Babenroth, Adolph Charles, and Viets, Howard T. , eds. Read-
        ings in Modern Business Literature. New York: Prentice-
        Hall, 1928. Bib. 595p.
This collection of readings is divided into two parts: "The New

Profession--Business, " and "Business Writing: Principles and Practices. " Each part is further subdivided into units (A, B, C, etc.),
which are further subdivided into 60 readings. Unit topics include
"Advertising as a Profession, " "Style, " "Selling by Mail, " "Public
Opinion, " "Modern Business and Civilization. " Authors in the collection include John Henry Cardinal Newman, John Ruskin, George
Burton Hotchkiss, and an editorial from the New York Herald Trib-
une. Included in the collection are two essays showing contrasting
views of business education. One essay advocates a solid business
education in colleges with no humanities; the other advocates a
liberal-arts background for business students so they can succeed
in the business of living. Each reading ends in discussion questions.

32 Baker, C. E. Foreign Commercial Correspondent. New York:
        Van Nostrand, 1901.

33 Baker, Clifford. A Guide to Technical Writing. New York:
        Pitman, 1961. Index. Bib. 101p.
This book is a ready source of reference "for either the specialist
publisher or the busy technician" and the practicing technical writer.
The first four of the 12 chapters deal with communication, technical
writing forms, the technical writer, and the audience. Chapters 5-8
treat grammar and usage and grammar problems, paragraph division,
readability, "mental planning, " and technical illustration. The remaining chapters cover editing, preparation of copy for publication,
reproduction techniques, and dissemination of technical papers within
and outside a company. The writing exercises primarily ask the
user to rewrite passages.

34 Baker, Clifford. Technical Publications: Their Purpose, Prep-
        aration and Production. London: Chapman and Hall,
        1955. Index. 302p.
This book is "a guide for the technician to the techniques of presenting information and producing it in the best--and cheapest--form. "
Chapter 1, "Forms of Technical Publicity, " is a description of various forms of technical writing. Chapter 2, "Meeting the User's
Needs, " focuses on audience analysis by defining the needs of typical
users of technical literature. Chapter 3, "The Use of Words, " covers usage, punctuation, and mechanics. Chapter 4, "Copy Writing, "
treats the writing process for technical documents. Chapters 5 and
6 cover the value and use of illustrations, and Chapters 7 and 8
cover printing, production, and copy-editing. Chapter 9 shows the
end products of technical publications in the aircraft industry. Chapter 10, "Legal Documents, " covers the legal writing factors in specifications and patents. The book is geared to British technical
publications.

35 Baker, Josephine Turck. Correct Business Letter Writing and
        Business English. Chicago: Correct English, 1911.
        205p.
The author stresses that a "business man's letter betokens either
his illiteracy or his culture. " This reference book is divided into
16 sections. Each section is illustrated with models. Topics in-

clude advice for addressing individuals, firms, clergy, married wom-
en, unmarried women, government officials. Also covered are punctuation rules, geographical names, abbreviations, and organization of
the letter. The last section is a 38-page list of compound words.

36 Baker, Ray Palmer, and Howell, Almonte Charles. The Prep-
         aration of Reports. rev. ed. New York: Ronald, 1938.
         Index. Bib. 578p.
The first edition of this book, published in 1924 with Ray Palmer
Baker as the only author, has been considered as a landmark in
technical writing textbooks. Although the details are dated, the
philosophy of this book sounds strikingly modern: "A report must
be so built as to carry the right view of a matter perhaps highly
technical in nature to readers who have no first-hand knowledge of
the case. " The scope of coverage in this text is very broad; it covers every conceivable report type in addition to writing characteristics and numerous samples of reports. Chapter 1, "Origins and
Forms, " even reviews the history of report writing, as far back as
Frontinous of Rome. Chapter 2, "Elements, " covers standard parts
of reports, and Chapter 3, "Characteristics, " covers the writing
style found in reports. Chapters 4-21 cover a wide variety of report types--including public, statistical, examination, recommendation, research, and theoretical reports. Chapter 22, "Preparation
of Manuscript and Revision of Proof, " treats formal editing and manuscript preparation as well as production techniques and copyright
issues. An appendix includes comments from academics, business
people, and engineers about the importance of report writing. Following the appendix is a bibliography of reports, books, and articles
on report writing.

37 Ball, John, and Williams, Cecil B. Report Writing. New York:
         Ronald, 1955. Index. Bib. 407p.
This textbook is divided into four parts: "Reports and Report Writing, " "Preparing the Report, " "Supplementary Readings and Special
Applications, " and "Work Materials. " The first 11 of the 18 chapters in Parts I and II provide instruction for writing, e. g. , "Style, "
"Visual Aids, " "The Formal Report. " Part III is a collection of
readings on report writing. Part IV contains four case studies of
reports. Included are topics for reports. The bibliography is divided into subjects, such as finance, management, and foreign trade,
with references for obtaining information in those fields. Each of
the first 11 chapters on report writing ends in exercises--either
writing assignments or class discussion.

38 Bamburgh, W. C. Talks on Business Correspondence. New
         York: Little and Ives, 1916.

39 Bancroft, John C. Writing That Sells. San Rafael, Calif.:
         Bancroft, 1975.

40 Banks, Eleanora. Putnam's Correspondence Handbook. 4th ed.
         New York: Putnam's, 1919. Index. 259p.
This book was originally published in 1912 under the title Correct

Business and Legal Forms by the same author, "with new material
added" in the 1914 and 1919 editions. This edition is a reference
book for those who write and edit letters. The 28 unnumbered sections primarily cover various aspects of form, punctuation, grammar,
and other mechanics. However, some advice on writing strategy is
included in addition to the model letters that are shown throughout
the book. No sections are devoted to special types of letters (claim,
adjustment, etc.). In addition to the business forms illustrated, the
book includes format advice for such documents as court testimony,
legal papers and wills, telegrams and cablegrams, and even poetry.
Other sections cover Latin words and phrases, general usage advice,
word choice, spelling, and proofreading. Some of the subjects have
a surprisingly modern flavor. In addressing letters to women, for
example, the book recommends omitting the salutation altogether as
an option--and the author laments "to make the matter more complicated, authorities do not wholly agree" on the best solution.

41 Banks, J. G. Persuasive Technical Writing. New York: Per-
       gamon, 1966.

42 Bar, Ulrich, ed. Encyclopedia of Business Letters in Four
       Languages. 3 vols. New York: Arco, 1971. Index.
       Vol. 1, 325p.; Vol. 2, 195p.; Vol. 3, 217p.
These volumes contain sample business letters in English, Spanish,
French, and German. The editor has "reproduced the meaning and
content" of the letters rather than given verbatim translations. The
letters include models for requests for references, orders, replies
to orders, requests for credit, opening an account, and stopping payment. The section "Interchangeable Sentences" contains alternatives
to some of the sentences in the model letters. The first volume
contains a glossary of international trade terms and abbreviations,
metric-conversion tables, a list of countries and their languages, a
list of countries and their currencies, and a glossary of postal and
transportation terms.

43 Barnes, Nathaniel Waring. How to Teach Business Correspond-
       ence. Chicago: Shaw, 1916. Bib. 83p.
This book is probably one of the earliest pedagogical works in business writing. It is divided into three sections: "Methods," "Materials," and "Assignments." In Part I, the author covers in two
chapters an overview of instruction in correspondence and various
approaches to use in the classroom. In Part II, the author provides
one chapter of readings about the principles of effective business letters, one chapter of typical letters and more articles about correspondence, and a bibliography of business methods and letter writing.
Part III contains three chapters of classroom assignments: general
information tests, writing exercises, and examinations for the chapters of How to Write Business Letters, a textbook by the same author.

44 Barnes, Nathaniel Waring. How to Write Business Letters.
       Chicago: Shaw, 1916.

45   Barnett, Marva T.   Elements of Technical Writing.   Albany,
          N.Y.:   Delmar,   1974.   Index.   232p.
This textbook is designed to "help technicians, professionals, and
administrators communicate knowledge of their specialized skills to
other interested personnel." Its 28 units are divided into five sections: "Guidelines to Technical Writing," "Planning the Report,"
"Writing the Report," "Report Supplements," and "Writing Letters."
The exercises include writing assignments and sentence/paragraph
correction problems.   The book also has a small section on the oral
report.

46   Barnett, Marva T., and Smith, J. L.   Effective Communications
          for Public Safety Personnel.   New York: Delmar,   1978.
          Index.   279p.
This textbook is aimed at students in police- and fire-training programs.   Its 26 units are divided into seven sections.   Section I,
"Introduction to Effective Communication," covers importance and
writer attitude.   Section II, "Correct Use of the English Language,"
explains grammar, punctuation, numbers, etc.   Section III, "Aids to
Effective Communication," deals with clarity, conciseness, and graphics.   Sections IV and V deal with informal and formal reports, including a unit on the officer's personal notebook.   Section VI covers
letters, and Section VII covers "Verbal and Nonverbal Communication," including court testimony and body language.   Each unit ends
in student exercises.   Unit 13, "Form Reports," contains illustrations of fire and police forms.

47   Barr, Doris W.   Communication for Business, Professional,
          and Technical Students.   2nd ed.   Belmont, Calif.:   Wads-
          worth, 1980.   Index.   495p.
Originally titled Writing, Listening, Speaking for Business and Professional Students, this textbook-workbook is designed for the vocational student.   The 15 chapters are divided into five parts.   Part
I covers the job search, and Part II covers the on-the-job communication forms: memos, workorder forms, letters, reports, etc.
Part III devotes eight chapters to writing patterns, such as narration, description, comparison, exemplification, and persuasion for
both writing and speaking.   Part IV covers writing business letters,
listening, and summarizing.   Part V deals with interviewing, interpreting data, questionnaires, and documentation.   A final 96-page
"Technical Review" provides grammar and sentence exercises.   Chapters end in sentence-correction exercises, writing assignments, and
exercises to complete standard forms.

48   Barr, Doris W.   Effective English for the Career Student.   Bel-
          mont, Calif.:   Wadsworth, 1971.   Index.   193p.
This book results from "decisions" that the author and her students
reached in career-communications courses.   The text uses a "hands-
on approach" to teaching writing.   Chapter 1, "Writing a Report,"
discusses the writing process, outlining, charts and tables, paragraphs, etc.   Interspersed throughout the chapter are writing as-
ments.   Chapter 2, "Casebook," contains ten cases requiring re-

sponses, such as a comparative report on typewriters, a process report, and a spatial description. Chapter 3, "Grammar and Style," provides 141 pages on grammar and punctuation. Exercises follow each topic covered.

49 Barrass, Robert. Scientists Must Write: A Guide to Better Writing for Scientists, Engineers and Students. London: Chapman and Hall, 1978. Index. Bib. 176p.

This book is designed primarily for scientists and engineers and secondarily for students in those fields. A few exercises are included so that the "book may be read either as an alternative to a formal course on scientific and technical writing or to complement such a course." The first three chapters discuss why writing is important to scientists and what they must write. Chapter 4, "How Scientists Should Write," gives the characteristics of good scientific writing. Chapter 5, "Think-Plan-Write-Revise," presents an overview of the writing process. The next three chapters cover word choice and making writing readable. Chapters on using numbers and using illustrations follow. Chapter 11, "Reading," treats the kind of material scientists should be reading and how reading is useful for scientific writing. Chapters 12 and 13 discuss the parts of reports and theses. Chapter 14 covers preparing a speech on a scientific subject.

50 Barrett, Charles R. Business English and Correspondence. 2nd ed. Chicago: American Technical Society, 1919. 205p.

The first edition of this text was published in 1914. It is divided into two parts, each with separate paging. Part I (102 pages), "Business English," consists of both explanation and exercises dealing with grammar and punctuation. Part II (103 pages), "Business Correspondence," covers style, paragraphing, openings, complaints, recommendations, credits, etc. All sections end in exercises.

51 Barry, Robert E. Business English for the '70's: A Text-Workbook. 2nd ed. Englewood Cliffs, N.J.: Prentice-Hall, 1975. Index. 386p.

This book emphasizes grammar, usage, punctuation, and mechanics. It is designed in an $8\frac{1}{2}$ " x 11" format with tear-out pages. The final eight of the 40 chapters serve as a handbook, including glossaries of terms used in business and general vocabulary words. The exercises are primarily fill-in and cross-out correction of sentences. A unique feature of the teaching apparatus is the use of a grammatical crossword puzzle. A 1980 edition has been published.

52 Bartholomew, Wallace Edgar, and Hurlbut, Floyd. The Business Man's English. rev. ed. New York: Macmillan, 1931. Index. 357p.

This text, originally written in 1920, "aims to develop an understanding of the English of business through the introduction of an unusually large number of exercises that deal with English as it is used in business situations." The first four chapters discuss business writing in general. Chapters 5 and 6 cover speaking and pronunciation. Chapter 7 through 9 cover word choice, sentence con-

struction for clarity and emphasis, paragraphs, and punctuation.
Chapter 12 treats letter format, and Chapters 13-19 cover various
types of letters (e.g., collection and sales) and form letters. Chapter 20 treats supervision of letter writing, and Chapters 21 and 22
cover the job search. The final seven chapters cover such topics
as report writing, parliamentary procedure, filing and indexing, and
postal information. For an early book, this textbook contains an
abundance of exercises, both at the end of the chapters and in the
appendix.

53 Barton, Everett H. Communication Requirements for Technical
        Occupations. Olympia: State of Washington Coordinating
        Council for Occupational Education, 1970. 39p.
This study focuses on communication skills and training needed for
vocational education in the state of Washington. After a brief overview of the purpose of this study and an accompanying chart showing
the communication skills covered (listening, talking, graphic interpretation, graphic presentations, reading, and writing), the report
provides a series of "Function Analysis Forms." These forms define the name of a function, such as "Write Job Instruction and Work
Orders," a definition of the function, and a "Rationale" (or description of the requirements of and effective job for the function). The
sections on writing, speaking, and listening may help teachers in
writing behavioral objectives.

54 Bates, Jefferson D. Writing with Precision: How to Write So
        That You Cannot Possibly Be Misunderstood; Zero Base
        Gobbledygook. Washington, D.C.: Acropolis, 1978.
        Index. Bib. 308p.
This self-teaching book with questions and answers is designed for
government writers. The book has a personal flavor and wages war
against gobbledygook in bureaucratic prose. It is divided into two
parts. Part I is divided into 14 study units on such principles as,
use the active voice, don't make nouns out of verbs, use parallelism, and arrange your material logically. Part II is the "Handbook,"
which is an alphabetically arranged list of definitions and short discussions of concepts that the author believes writers need. Also
included is a bibliography that lists a wide variety of style books,
dictionaries, word-finders, and other guides. Additional questions,
exercises, and answers follow the two parts.

55 Beardwood, Lynette; Templeton, Hugh; and Webber, Martin.
        A First Course in Technical English--Book I. London:
        Heinemann Educational Books, 1978. 103p.
This textbook is for beginning students in English who need the language for their technical studies. There are eight units dealing with
such topics as geometrical shapes, instructions, and electricity.
Each unit provides three readings followed by questions and exercises. Tape-recordings of the readings are available for pronunciation practice, vocabulary, and listening comprehension.

56 Beardwood, Lynette; Templeton, Hugh; and Webber, Martin.
        A First Course in Technical Writing--Book II. London:

Heinemann Educational Books, 1979.   83p.
This textbook is a continuation of the authors' Book I.   Book II provides practice for foreign students in English at the intermediate
level.   The eight units deal with such topics as steam engines, industrial chemistry, and welding.   Each unit has three readings followed by questions and exercises.   Tape-recordings of the readings
are available to provide practice in speaking and listening.

57 Belding, Albert G.   Business Correspondence and Procedure.
          New York: Ronald, 1922.   Index.   397p.
The twenty-six chapters in this textbook are aimed at teaching business writing "in the modern way."   Chapter topics include the function and cost of letters, parts of the letter and stationery, types of
letters, business style, spelling, grammar and punctuation, usage,
abbreviations, proofreading, form letters, filing, and contracts.
Chapter 17 presents guidelines for "Your First Office Job"--primarily
tips on sorting mail and office routine.   Chapters 24-26 discuss effective use of the telephone and telegram.   Appendix A offers writing
assignments and case studies.   Appendix B contains copies of "recent examinations by the University of the State of New York, high
school division" on the subject of business correspondence.

58 Bell, Reginald William.   Write What You Mean.   Foreword by
          Sir Ernest Gowers.   London: Allen and Unwin, 1954.
          Index.   116p.
This book is geared to the businessperson seeking advice on clear
business writing.   The author stresses the problem of preparing to
write.   The first two chapters cover gathering material, selecting
evidence, and formulating conclusions.   Chapter 3 gives general advice on document style and layout.   Chapter 4 covers tone and in-
tention.   Chapters 5 and 6 cover writing and answering letters.
Chapters 7 and 8 apply the writing principles to reports, memoranda,
minutes, orders, regulations, and announcements.   In Chapter 9, the
preparation of documents for court is discussed.   The appendix,
"Ten Reminders on Use of Language," lists tips for simplicity and
and provides a paragraph of illustration for each tip.

59 Bender, James F.   Make Your Business Letters Make Friends.
          New York: McGraw-Hill, 1952.   Index.   250p.
A product of the author's course "Communication for Executives,"
this book is designed to help business people write better, more
friendly letters.   It contains tips for writing letters, brief questionnaires about the reader's writing habits, and lists of words to use
or avoid.   In addition to chapters containing traditional advice, some
of the 12 chapters present samples of effective letters, memos, and
notes; advice for dictation; and diagnostic tests for writing ability.
Other than the diagnostic tests, no exercises are included.

60 Bennett, John Barnard.   Editing for Engineers.   New York:
          Wiley-Interscience, 1970.   Index.   Bib.   126p.
Although this book is designed to help engineers who write documents
and the managers who edit them (no exercises are included), advanced
students of technical writing could use it to study the entire process

of editing and manuscript preparation.   Each of the six chapters begins with an abstract.   The topics covered include the definition of
editing and the manager's role, working with the writer, editor as
a teacher, editing guidelines, support personnel to an editor, and
documentation systems for companies.   Following the chapters are
ten reference appendixes, including "Readability," "Reference Books,"
"Indexing and Abstracting Techniques," "Grammar," "Punctuation,"
"Compound Words," "Capitalization," "Numerals," "Sample Style
Guide," and "Duplication Process."

61   Berenson, Conrad, and Colton, Raymond.   Research and Report
          Writing for Business and Economics.   New York: Ran-
          dom House, 1971.   Index.   Bib.   182p.
This book is a style guide for students, scholars, and those who
must write reports and research papers in business fields.   Its 17
chapters begin with an overview of the research process and methods
and move to specific topics, such as format, footnotes, and organization of parts.

62   Berman, Alan J., and Feder, Irwin.   Basic Business Communi-
          cation, Writing Your Way to a Successful Career.   New
          York: Pella, 1978.   240p.
This textbook grew out of a course given by the authors at LaGuardia
Community College of the City University of New York.   The book is
in three parts.   Part I, "First Things First: Landing a Job Through
Writing," covers the job search, application, résumé and follow-up
letter.   Part II, "You're There Now!:   Some Basic Types of On-the-
Job Writing," covers types of letters (inquiries, adjustment, collection, sales), memos, and short reports.   Part III, "Appendices,"
contains a discussion of effective writing principles (clarity, conciseness, etc.), the job interview, taking minutes, a glossary of 101
useful business terms, and a list of some business periodicals.
There are numerous checklists, along with discussion questions and
writing assignments.

63   Berset, Francis.   Commercial Correspondence in Four Lan-
          guages.   Biel, Switzerland: Edition Du Panorama, 1959.
          Index.   237p.
This book is composed of business letters, printed in side-by-side
columns, in four languages: English, French, Spanish, and German.
It is useful for someone who wishes to write a letter in another language without learning the language.   The user simply finds the letter
type in the index (also in four languages), goes to the page indicated,
then copies the letter in the other language.   Thirty-six letter types
are listed in the index.   The book contains no introduction, overview, or tips on composing letters.

64   Better Business Relations Through Letters to Employees.   Wash-
          ington, D.C.:   Chamber of Commerce of the United States
          of America, 1965.

65   Better Letters: A Little Book of Suggestions and Information
          About Business Correspondence.   Chicago: Browne, 1920.

66 Bishop, Malden Grange. Billions for Confusion: The Technical
        Writing Industry. Charlotte, N.C.: McNally and Loftin,
        1963. 123p.
This book is intended to inform the public about military spending
on contracts for writing technical information on the products supplied to them.   The book also gives tips to the prospective technical
writer on how to communicate technical information effectively--
especially about not forgetting the reader.   The nine chapters of the
book deal with such areas as the causes of confusion and the break-
down of communication, how to avoid pitfalls in writing, and how
successfully to meet the challenge of competent communication.   The
book is full of anecdotes from the author's experience with the military, exposing the blunders that the Air Force, Navy, and Army
have committed in allowing an overflow of incomprehensible documents.   The book also gives ideas on reducing the enormous cost
of preparation and dissemination of technical publications.

67 Bishop, Malden Grange. Go Write, Young Man. Three Rivers,
        Calif.: Del Malanbob, 1961. 135p.
This book is a probing discussion of the field of technical writing--
what it is, responsibilities of the writer, professionalism, and problems encountered.   The discussion is pertinent to those entering
the field or explaining the field to colleagues.   There are six chapters discussing the writer and "the challenge," "qualifications,"
"writing," "the customer," "professionalism," and "The Writer as
a Professional." Bishop warns, in particular, that technical writers
must be professionals--only then will there be no conflict or overlap
with other fields.

68 Bithell, J. Handbook of German Commercial Correspondence.
        New York: Longmans, Green, 1908.

69 Blacke, David. English for Basic Maths. Sunbury-on-Thames,
        England: Nelson, 1978.

70 Blickle, Margaret D., and Houp, Kenneth W. Reports for Sci-
        ence and Industry. New York: Holt, 1958. Index.
        320p.
This textbook of 11 chapters covers such topics as varieties of reports, format, business letters, proposals, memoranda, and progress
reports.   Chapter 10, "The Informative Article," covers markets for
articles as well as writing the articles.   Chapter 11, "Language,
Composition, and Writing Mechanics," treats organization, paragraphing, and grammar.   All chapters end in discussion questions and
writing assignments.

71 Blickle, Margaret D., and Passe, Martha E., eds. Readings
        for Technical Writers. New York: Ronald, 1963. 367p.
This anthology is divided into three parts.   Part I, "Language and
Usage," covers some basic definitions of technical and business
writing and their styles.   Part II, "A Range of Style," includes articles by science writers on scientific subjects.   Part III, "A Range
of Subjects," includes articles ranging from general discussions

about science to definitions of various fields in science.   Each article
is followed by a set of discussion questions.   Each of the three parts
is preceded by a brief introduction.   Articles include "What Is Technical Writing," by Robert Hays, "Poetic and Scientific Discourse,"
by Cleanth Brooks and Robert Penn Warren, "Soap Bubbles," by
Charles V. Boys, and "A Humanist Looks at Science," by Howard
Mumford Jones.   Most of the articles were published at least ten
years prior to the date of this anthology.

72 Blicq, Ron.   On the Move: Communication for Employees.
         Englewood Cliffs, N. J.:   Prentice-Hall, 1976.   Index.
         259p.
This text is for "people on the move ... upward within a company
and, sometimes, outward into the world of small business owner-
ship."   There are six parts: "This Business Called Communication,"
"Communicating with Prospective Employers," "Communicating as
an Employee," "Communicating When You Become Supervisor,"
"Communicating When You Go into Business for Yourself," and "The
Style and Shape of Written Communications."   Topics covered, as
the reader advances from the job search to owning his or her own
business, include communication theory; the job search; making suggestions; calming the dissatisfied customer; listening to employees,
giving instructions, and communicating at meetings; reporting progress, investigations, and events; quoting prices and billing; hiring;
writing for a reader; and word choice.   Exercises give case situations where the reader responds either in spoken or written form.
The instructor is to assign someone to act as receiver to the spoken
answers.   The book is full of tips on the politics of employer-employee
and employee-customer communications.

73 Blicq, R. S.   Technically--Write!   Communication for the Tech-
         nical Man.   Englewood Cliffs, N. J.:   Prentice-Hall, 1972.
         Index.   385p.
Designed for the technician, technologist, engineer, or scientist, this
book is divided into 11 chapters covering such topics as reports,
technical correspondence, and other forms of technical writing (e. g.,
instructions, parts lists, papers, and articles).   Also covered are
using illustrations, giving oral presentations, and revising sentences.
Some chapters end with cases to which the reader responds by writing various forms.   Following the chapters are sections on punctuation, grammar, and a glossary of technical usage.

74 Block, Jack, and Labonville, Joe.   English Skills for Techni-
         cians.   New York: McGraw-Hill, 1971.   Index.   Bib.
         210p.
This text consists of eight chapters: "Introduction," "Purpose,"
"Content," "Organization," "Written Reports," "Oral Reports," "Getting a Job," and "Grammar."   The sections within the chapter are
numbered in military style.   Each chapter ends in long fill-in writing exercises ranging from describing an automobile to identifying
nouns and verbs.   Chapter 7, "Getting a Job," contains a list of
94 interview questions compiled from a 1964 study.

75 Blumenthal, Lassor A.   The Complete Book of Personal Letter-
            Writing and Modern Correspondence.   Garden City, N.Y.:
            Doubleday, 1969.   Index.   313p.
This book deals with not only the usual range of business letters
(e. g., claims, inquiries, job applications) but also with a wide variety
of personal letter-writing situations (e. g., "Birth Announcements,"
"Thank-You Notes," "Christenings"), "Tenant-Landlord Letters,"
"Writing for Club Members," and "Writing Classified Ads." The 11
chapters of the book give numerous examples of all these types of
writing.   The book constantly stresses that letters should be written
in the appropriate and socially acceptable form to bring about the
desired responses.   The book includes "guidelines, checklists, and
hundreds of model letters." There are no exercises.

76 Blumenthal, Lassor A.   Successful Business Writing: How to
            Write Effective Letters, Proposals, Résumés, Speeches.
            New York: Grosset and Dunlap, 1976.   Index.   Bib.
            112p.
The first two chapters deal with organizing ideas logically and writing
without errors in grammar and mechanics.   Chapters 3-9 deal with
different types of correspondence, such as letters of inquiry, collection, and complaint.   Chapters 10-15 suggest ways of writing effective memos, proposals, and résumés, as well as making oral presentations.   The last three chapters cover letter format.   The book
also gives practical advice on selecting stationery and letterheads
and provides numerous models of the forms of business writing.

77 Bogard, Morris R.   The Manager's Style Book: Communication
            Skills to Improve Your Performance.   Englewood Cliffs,
            N.J.:   Prentice-Hall, 1979.

78 Bonner, William H.   Better Business Writing.   Homewood, Ill.:
            Irwin, 1974.   Index.   468p.
Since this textbook is designed for the beginning college student, the
first four chapters cover the principles of communication and the
format and style for business writing.   These principles are applied
to three business forms--letters, formal reports, and informal re-
ports--which are treated extensively in ten chapters of the book.
One chapter each is devoted to "Dictating," "Communicating Orally,"
and "Using Graphic Aids." Included are many samples of business
letters, and a large number of questions, problems, and exercises
are provided at the end of every chapter.   A 68-page reference section at the end of the book covers punctuation, documentation, letter
format, memos, and interview suggestions.

79 Bonner, William H.   Communicating Clearly: The Effective
            Message.   Chicago: Science Research Associates, 1980.
            Index.   233p.
This workbook is designed to help students develop proficiency in
English for writing on the job.   The 23 chapters are divided into
five parts.   Part I covers word choice, the dictionary, spelling, and
mechanics.   Part II explains parts of speech; Part III treats punctuation, abbreviations, and end-of-line word division.   Part IV covers

sentences and paragraphs, and Part V deals with letters, memos, and envelope formats. Students have write-in exercises and writing assignments.

80 Bonner, William H., and Voyles, Jean. Communicating in
        Business: Key to Success. Houston: Dame, 1980. In-
        dex. 388p.
This textbook has 17 chapters covering communication principles, types of letters, format, public-relations messages, employment messages, forms and form letters, dictation, graphics, reports, and oral communication. The 54-page reference section covers punctuation, selection of stationery, and letter and report format. Each chapter ends in discussion questions, writing assignments, and case problems.

81 Bonney, Louise E., and Cole, Carolyn Percy. Handbook for
        Business Letter Writers. New York: Harcourt, Brace,
        1922. Index. Bib. 98p.
The authors, high school teachers, designed this book "to teach young people to handle intelligently those essential tools of business, words and sentences." The book is divided into 13 sections following a foreword about why business letters should make friends. The sections cover letter form, style, diction, grammar, punctuation, use of numbers, abbreviations, spelling, syllabication, types of letters, and telegrams. No exercises are included.

82 Boone, Anne. Modern Business Letter Writing. New York:
        Ronald, 1937. Index. 251p.
This book covers "how to use the words and phrases that count; how to make letters concise but complete; how to develop speed in handling correspondence efficiently." The first nine of the 28 chapters deal with the tools of clear and concise writing; avoidance of "Jargon"; "Faulty Expressions"; "Completeness"; and "Promptness." Chapters 10 and 11 treat the openings and closings of letters. Chapters 12-22 deal with appropriate tone, courtesy, and the "you attitude" displayed in business correspondence. Chapters 23-28 treat word choice, "Idiom," "Specific vs. General Language," and "Paragraphing." Chapter 24 contains two lists: (1) words reinforcing positive effect and (2) words showing negative attitudes. A frequent device is to show actual business letters before and after revision. There are no exercises, but a great many examples are included to illustrate stylistic and psychological principles of competent business writing.

83 Bosticco, Mary. Instant Business Letters. London: Business
        Books, 1968. 198p.
This book aims to improve the quality of and reduce the time it takes to write business letters. The book is primarily an anthology of canned openings, closings, and bodies for various letter situations. The writer is advised to make a pile of letters requiring a routine reply, select suitable paragraphs from the book, and jot down their numbers on the letter to be answered. The typist can then type out a letter including pertinent details for this particular letter. Part I,

"The Business Letter," covers appearance, addresses, effective
usage.   Part II, "The Instant Letter-Writer," contains the anthology
of canned parts.   Parts are included for sales, credit, collection,
personnel, orders, requests, and announcements.

84 Bovee, Courtland L.   Better Business Writing for Bigger Profits.
        Jerico, N. Y. :   Exposition, 1970.   238p.
Aimed at business people, this collection of 27 articles by various
authors is designed to "help you save time and money in using the
indispensable tools of business--letters and memoranda." The first
articles cover the business letter; the final articles cover the memo.
Although the author provides no commentary, some of the articles
are unique.   Two examples are "An Index of Insults for Writers of
Business Letters," by J. Harold Janis, and "Producing an Effective
Business Letter Manual," from Changing Times.

85 Bovee, Courtland L.   Business Writing Workshop: A Study Guide
        of Reading and Exercises.   Dubuque, Iowa: Kendall/Hunt,
        1976.   225p.
This anthology contains 41 readings divided into nine parts: "Writing
Effectively in Business," "Handling Inquiries and Acknowledgments
and Building Goodwill," "Getting Letters to Sell," "Responding to
Consumer Claims and Requests for Adjustments," "Special Types of
Business Letters," "Getting a Job," "Producing Memos," and "Writ-
ing Business Reports."   The articles include several by the author,
"How to Say It with Statistics," by Rudolf Flesch; "Liberate Your
Sales Letters," by Luther A. Brock; and "Writing Reports That Lead
to Effective Decisions" by Norman Sigband.   After each article, there
are four or five discussion questions.   The book contains no general
introduction or openings to the parts.

86 Bovee, Courtland L.   Techniques of Writing Business Letters,
        Memos, and Reports.   Sherman Oaks, Calif. :   Banner
        Books International, 1978.   90p.
This textbook is divided into six parts.   Part I, "Principles of Written Communication," discusses the importance of clear writing.   Part
II, "Principles of Writing Letters," covers various types of letters
and effective tone.   Part III, "Principles of Writing Memoranda,"
discusses in two pages the purpose of the memo.   Part IV, "Principles of Writing Formal Reports," covers the types of reports,
components of the long report, and format.   Part V, "Principles of
Writing Informal Reports," explains in two pages the difference between the formal and informal report.   Part VI, "Appendixes," contains "References for Business Writers," a short list of books and
journals; "Skill-Building Writing Exercises," 25 exercises for sentence revision or word choice; and "Checklist of Wordy and Trite
Expressions," with alternatives to 221 of these expressions.

87 Bovee, Courtland L.   Techniques of Writing for Business, In-
        dustry, and Government.   San Diego: Presidio, 1972.

88 Bowen, Mary Elizabeth, and Mazzeo, Joseph A. , eds.   Writing
        About Science.   New York: Oxford University Press,

1979. 353p.
According to the authors, "this anthology aims to demonstrate the
range, the rhetorical complexity, and the sheer excitement of scientific writing. . . . " The 26 articles, written by such authors as Isaac
Asimov, Lewis Thomas, and Rachel Carson, are divided into two
parts: "Writing for Popular Audiences, " and "Writing for Professional Audiences. " A rhetorical table of contents is included, listing
the articles under such topics as "Definition, " "Comparison and Con-
trast, " and "Analysis of Cause and Effect. " Each reading ends with
study questions and suggestions for writing.

89 Bowers, Warner Fremont.   Techniques in Medical Communica-
           tion. Springfield, Ill.: Thomas, 1963. Index. 88p.
This book, addressed to the medical professional, places its emphasis more on writing than on speaking. The first of the five sections
treats public speaking with outlines and visual aids. The second section treats medical writing, the problems of the author, and those of
the editor. The third section covers "Methods of Data Presentation";
use of statistics; collection, analysis, and presentation of data; and
the construction of lantern slides. The fourth section treats the
management of the medical meeting, scheduling of events, personnel
requirements, program planning, finances, and public relations. The
final section is devoted to the preparation of illustrations.

90 Bowman, Joel P. , and Branchaw, Bernadine P.   Successful
           Communication in Business. San Francisco: Harper and
           Row, 1980. Index. Bib. 593p.
The 20 chapters of this textbook are divided into five parts. Part I
covers communication theory, reader and purpose, major elements of
business communication (clarity, courtesy, and tone), problem solving, and format.   Part II covers typical messages: orders and requests, positive and negative messages, persuasive messages, and
writing messages with more than one purpose.   Part III focuses on
business reports: types, organization, and oral reports.   Part IV
explains the job-application process.   Part V covers organizational
communication and techniques for self-improvement (e. g. , listening,
reading, memory development).   All chapters end with writing and
case problems.   The four appendixes include grammar review, sample reports, additional case problems, and correction symbols.

91 Bowman, Joel P. , and Branchaw, Bernadine P.   Understanding
           and Using Communication in Business. New York: Harp-
           er and Row, 1977. Index. Bib. 275p.
Originally designed as a core for satellite books on specific business
communication applications, this book contains 12 chapters on such
topics as communication theory, writing skills, communication skills
(listening, reading, group, nonverbal, dyadic, interpersonal), organizational communication, and job application.   The chapters begin with
behavioral objectives and conclude with chapter summaries and exer-
cises--discussion questions, correction of sentences, and writing assignments.

92 Boyd, William P. , and Lesikar, Raymond V.   Productive Busi-

ness Writing. Englewood Cliffs, N. J.: Prentice-Hall,
1959. Index. 513p.
The primary emphasis of this textbook is on writing letters in business; however, one chapter on reports is provided. Chapter 1 covers
basic principles of letter writing. Succeeding chapters cover letter
appearance, inquiry and request letters, orders and acknowledgments,
sales letters, adjustment letters, credit letters, collection letters,
and job applications. The final two of the 11 chapters treat dictation
and speed in letter writing, and grammar and punctuation in letters.
A special feature of this book is the inclusion of "grading check
lists" for the various types of letters so that the students and teachers have a common standard for evaluating performance. All exercises are case problems.

93 Brand, Norman, and White, John O. Legal Writing: The Strat-
        egy of Persuasion. New York: St. Martin's, 1976. In-
        dex. 205p.
This textbook, "designed for the prelaw student, the paralegal student,
and the law student, " is meant for practice in clear and persuasive
argumentation. The first of the six chapters covers the general
principles of writing; the second two chapters treat the special techniques of legal exposition and persuasion. Chapters 4 and 5 deal
with the problems of sentence and paragraph structure and poor writing style, as they apply to the specific demands of legal writing.
The final chapter covers "Logic and Argument, " the core strategy
of legal writing. The chapters conclude with exercises for class
discussion and writing as well as case problems.

94 Brennan, Lawrence David. Business Communication. Paterson,
        N. J.: Littlefield, Adams, 1960. Index. Bib. 320p.
This book of 17 chapters covers the types of business-correspondence
situations. Each chapter ends with questions for class discussion and
with writing assignments. Topics include the importance of clear
communication, communication principles, letter appearance, organization, memos and inquiry letters, sales letters, promotional material, résumés, credit and collection letters, claims and adjustments,
and reports. One chapter covers business speaking. Chapters 14-16
cover spelling, grammar, and punctuation.

95 Brewer, John. Readings in Technical Report Writing. Colum-
        bia, Mo.: Lucas Brothers, 1959. Index. 164p.
According to the author, "the following readings have been written
under my supervision by the students of the Missouri School of Mines
and Metallurgy--students who studied technical report writing during
the school year of 1958-59. Each selection has been written as a
partial fulfillment of the engineer report writing course, English 51.
Each student listed as having prepared a report was assigned the
task of putting the essence of one of my lectures into good report
form. Each title is taken from our course syllabus. " The 51 student reports cover such topics as bibliography, footnotes, definition,
logical proof, argumentation, ethical proof, persuasion, semantics,
prefaces, outlining, inspection-trip reports, investigation reports,
denotation, research interviews, questionnaires, usage, connotations,
unnecessary words, and logical development of paragraphs.

96 Brock, Luther A.   How to Build Goodwill Through Credit Cor-
        respondence.   New York: National Association of Credit
        Management, 1976.

97 Brock, Luther A.   How to Communicate by Letter and Memo.
        New York: McGraw-Hill, 1974.   Index.   131p.
The author of this unusually styled textbook (folded in three panels
with a spiral-bound text and pad of exercises keyed to chapters) says
the design is intended to "streamline" the learning process.   The
eight chapters include "How a Business Letter Differs from an English Theme"; "Ingredients of Successful Business Letters"; "Memos
in Business: Tools for Management"; "Type 'A' Letters (good news);
"Type 'B' Letters" (bad news); and "Type 'C' Letters" (persuasive).
The appendix is a grammar review.   The exercise portion is keyed
to the text and includes questions based on the theory in the chapters, sample letters for evaluation, case problems, and sentence
exercises.   All writing is to be done in the workbook portion.   Chapter 8 gives tips on dictating letters and memos.

98 Brogan, John A.   Clear Technical Writing. New York: McGraw-
        Hill, 1973.   Bib.   213p.
The purpose of this programmed, self-study guide is to help both
professionals and students "develop a writing style that is direct,
clear, and concise. " The book's 16 chapters are divided into four
parts: "Removing Redundancies, " "Unleashing Verb Power, " "Using
Lean Words, " and "Stressing What Is Important. " The entire book
is in question-and-answer format; no sections are devoted entirely to
exercises.

99 Bromage, Mary C.   Cases in Written Communication II.   rev.
        ed.   Ann Arbor: University of Michigan Press, 1973.

100 Bromage, Mary C.   Writing for Business.   3rd ed.   Ann Ar-
        bor: University of Michigan Press, 1973.

101 Brown, Harry M.   Business Report Writing.   New York: Van
        Nostrand, 1980.   Index.   350p.
This textbook of 11 chapters is divided into three parts.   Part I,
"Principles of Business Communication, " covers the qualities of effective reports and reader analysis.   Part II, "Tools of Effective
Report Writing, " deals with paragraphs, sentences, word choice, examining evidence, and graphics.   Part III, "Writing Reports, " discusses the steps involved in writing a long report.   Each chapter
ends in writing assignments or exercises.   Appendix A is a glossary
of business terms; Appendix B covers common business abbreviations;
Appendix C is a glossary of general usage; and Appendix D covers
punctuation.

102 Brown, Harry M. , and Reid, Karen K.   Business Writing and
        Communication: Strategies and Application.   New York:
        Van Nostrand, 1979.   Index.   404p.
The 13 chapters of this textbook for basic courses are divided into
three parts.   Part I covers preparation for writing business letters,

letter format, accuracy, word choice, conciseness, clarity, the "you attitude, " and tone.   Part II explains kinds of communication: routine, favorable and unfavorable, sales, job applications, and reports. Part III consists of a single chapter on oral communication.   Chapters end with exercises, writing assignments, and cases.   The three appendixes cover forms of address, grammar, and usage.   A glossary of business terms follows.   Included is an index of the model letters in the chapters.

103 Brown, James.   Casebook for Technical Writers.   San Francisco: Wadsworth, 1961.   232p.
Designed as a supplement to a technical-writing textbook, this book contains 26 chapters in two sections.   Section I, "Problems, " consists of basic exercises in converting information into written form (e. g. , a diagram of a machine is shown and the student is to write a description).   Section II, "Cases, " provides "full-scale situations with many complicating elements of context that must be considered in the achievement of communication. "   With each case there are a dozen or so writing assignments, asking the student to prepare a working sketch of a machine part or prepare a table summarizing the drilling requirements for production of a part, or write a logical summary of operations and production facilities.   The 14 cases involve the building of a cannon, traffic-flow investigation, brake-control valve, an automatic shutoff valve, and the processes of a sulfuric-acid plant.

104 Brown, James.   Cases in Business Communication.   Belmont, Calif. : Wadsworth, 1962.   Index.   278p.
This casebook presents eight situations to which the student is asked to respond.   The first case requires a simple routine letter.   The final case (over 100 pages) requires a variety of responses (letters, reports, etc.) to complex problems at a hypothetical milk-products company.   Following each case, often illustrated with letters, a section titled, "Applied Assignments" asks the student to respond to that case.   Following this section, a series of "Related Assignments" provides further discussion questions and writing assignments.   Many of the cases are illustrated with tables and diagrams.   Following the eight cases is an "Assignment Index, " listing types of business writing assignments and cross-referencing that particular assignment within the eight cases.

105 Brown, Leland.   Communicating Facts and Ideas in Business. 2nd ed.   Englewood Cliffs, N. J. : Prentice-Hall, 1970. Index.   443p.
This textbook of 12 chapters covers such topics as "Creating That Favorable Impression, " "Adapting Language, " "Writing Persuasively, " "Gathering Facts for Decisions and Reports, " and "An Overview: Toward a Broader Perspective. "   Three appendixes cover dictation and telephone techniques and rules of grammar and punctuation. There are writing problems at the ends of chapters and interspersed throughout.   Chapters end in summaries and suggested references. Several chapters also contain review quizzes on grammar, punctuation, or bibliographic form.

106 Brown, Leland. Effective Business Report Writing. 3rd ed.
         Englewood Cliffs, N. J. : Prentice-Hall, 1973. Index.
         Bib. 449p.
This textbook is designed for the student and the person on the job.
Chapter 1 discusses the communication process related to reports;
Chapter 2, types of reports; Chapter 3, the writing process, from
the analysis of reader, problem, and situation to matters of writing
style, persuasion, and logic; Chapter 4, visual aids; Chapter 5, oral
reports; Chapters 6 and 7, gathering and using data; Chapter 8, revising and editing; Chapter 9, parts of a report; Chapter 10, employee/
employer relations through internal publications; Chapter 11, the annual report as a public-relations device. Three appendixes follow
the chapters: "Sources of Business Information, " "Suggestions for
Reference, " and "Bibliography of Selected References. " The exercises include both writing assignments and cases.

107 Brown, Stanley, ed. Business Executive's Handbook. 4th ed. ,
         rev. by Doris Lillian. New York: Prentice-Hall, 1953.
         Index. 1, 496p.
Each of the 12 sections in this massive volume has a detailed table
of contents. Three sections are directly related to writing: Section
I, "Successful Selling by Direct Mail"; Section V, "How to Write Effective Business Letters"; and Section X, "Credit and Collections. "
The topics covered in these sections include how to write sales letters, sales-letter campaigns, legal aspects of business letters, letters that build goodwill, how to write credit letters, how to use form
collection letters, and so on. The authors have published sections
of this handbook separately.

108 Brunner, Ingrid; Mathes, J. C. ; and Stevenson, Dwight W.
         The Technician as Writer: Preparing Technical Reports.
         Indianapolis: Bobbs-Merrill Educational, 1980. Index.
         231p.
Designed for the community-college and the two-year-college student,
this textbook emphasizes audience analysis and a systematic procedure for designing reports. The ten chapters cover audience, purpose, report structure and arrangement, writing and editing sentences
and paragraphs, format and visuals, writing process, and oral reports. Chapters end in correction exercises and writing assignments.
The textbook is typeset in double-column, heavy, dark print.

109 Brusaw, Charles T. , and Alred, Gerald J. Practical Writing:
         Composition for the Business and Technical World. Bos-
         ton: Allyn and Bacon, 1973. Index. Bib. 203p.
This textbook emphasizes a five-step writing process: preparation,
research, organization, writing, and revision. The 16 chapters are
divided into these steps with a sixth part on the "Application" of the
writing process. Part I, "Preparation, " provides an overview of
the writing process and considerations preliminary to writing: determining the reader's needs, the writer's objective, and the scope.
Part II, "Finding Information, " covers taking notes, using the library,
and using interviews and questionnaires. Part III, "Getting It Together, " treats methods of development, outlining, and using illus-

trations.   Part IV, "Writing," covers techniques for achieving emphasis and approaching the rough draft, and Part V, "Fine Tuning,"
treats various facets of revision for conciseness, clarity, and accuracy.   Part VI, "Application," briefly covers types of business
and technical writing.   Included are discussion questions, exercises,
and writing assignments.

110   Brusaw, Charles T.; Alred, Gerald J.; and Oliu, Walter E.
          The Business Writer's Handbook.   New York: St. Mar-
          tin's, 1976.   Index.   Bib.   575p.
Designed as a handbook supplement to a business writing text, this
work is also useful for professionals.   The contents of this handbook
are alphabetically arranged with over 500 entries covering usage
(e. g., activate vs. actuate), parts of speech (e. g., verbs, clauses,
pronouns), types of business writing (e. g., correspondence, house-
organ articles, annual reports), format and illustrations, writing and
rhetorical principles, and mechanics.   The handbook employs a four-
way access system: the alphabetical entries themselves, which contain cross-references within them; a 52-page index; a "Topical Key"
to the alphabetical entries; and a "Checklist of the Writing Process,"
which arranges key entries with page numbers in a five-step writing
process.   The front matter includes a section on "How to Use This
Book" and "Five Steps to Successful Writing."

111   Brusaw, Charles T.; Alred, Gerald J.; and Oliu, Walter E.
          Handbook of Technical Writing.   New York: St. Martin's,
          1976.   Index.   Bib.   571p.
Like the parallel edition, The Business Writer's Handbook, by the
same authors, this book is designed as a handbook supplement to a
traditional textbook for technical writing courses.   The handbook is
identical to its parallel edition with the exception that many of the
examples in the more than 500 entries are from technical writing
contexts.   The handbook includes technical writing forms, such as
specifications, government proposals, technical manuals, and laboratory reports.   For information on the arrangement and general scope
of this handbook, see the preceding annotation, for The Business
Writer's Handbook.

112   Buck, Charles Edgar.   The Business Letter-Writer's Manual.
          New York: Doubleday, Doran, 1924.   232p.
This book, designed for "business men" and students, contains writing assignments and cases at the end of many of its chapters.   Although the 20 chapters cover many standard types of letters and general advice on style (e. g., avoiding hackneyed expressions), they
also focus on specific parts of the letter (e. g., openings and closings).   The book also covers punctuation and grammar--of "special
value to the secretary and stenographer."   The final chapter provides
some samples of "good, average business letters."

113   Buckley, Earle A.   How to Increase Sales with Letters.   New
          York: McGraw-Hill, 1961.

114   Buckley, Earle A.   How to Write Better Business Letters.

New York: McGraw-Hill, 1936. Index. 185p.
This book is aimed at the letter writers "who could be good or even
outstanding letter writers with a little of the right kind of study and
application." Filled with illustrations, the 14 chapters cover such
topics as "The Formula for Sales Letters," "Writing the Opening,"
"Writing the Body," "Writing the Close," and "Pointers on Specific
Types of Letters." Chapter 14 is a 59-page discussion of "Dictated
Letters, Divided into Twenty-two Separate Discussions." The separate
discussions include conciseness, openings, closings, tone, and specific letter types. Chapter 7, "Letter Problems," contains five problems, which the author solves by writing appropriate letters and providing running commentary on audience and content.

115 Buehler, Mary Fran. Report Construction: A Handbook for
        the Preparation of Effective Reports. Sierra Madre,
        Calif.: Foothill, 1970. Bib. 57p.
This textbook is intended for college students in engineering or the
sciences. The 21 chapters give step-by-step advice on preparing a
technical report, stressing organization rather than writing style.
The topics are covered in the order the writer would normally need
them: "The Subject," "The Readers," "Technical Writing: What Is
It?" "The Outline Form," and "Writing the Rough Draft." Two appendixes are included: Appendix A, "The Greek Alphabet," and Appendix B, "Report Heading Systems and Other Mechanics." No exercises are provided.

116 Buhlig, Rose. Business English. rev. ed. Boston: Heath,
        1922. Index. 473p.
This textbook is offered as "a course that will more nearly meet the
public demand for effective English teaching in the high school." The
author also prepared a teacher's manual with daily assignments for
two years of classwork. The 22 chapters are divided into four parts:
"Word Study and Grammar" reviews parts of speech, spelling, and
pronunciation. "Oral English" covers speech to entertain, inform,
and persuade. "Correct Business Writing" deals with letters, punctuation, sentences, and dictation practice. "The Newspaper, The
Sales Letter, The Advertisement" concludes the volume. Each chapter consists primarily of exercises on the topic. The 452 exercises
vary from sentence correcting to writing brochures.

117 Burch, George Edward. Of Publishing Scientific Papers. New
        York: Grune and Stratton, 1954. 40p.
This essay "was presented as the Presidential Address at the meeting
of the Central Society for Clinical Research on October 30, 1953 in
Chicago." The essay defines what constitutes "a good scientific paper"; the characteristics of original research; sound interpretation of
the data; writing in readable, clear prose; and the author/editor relationship.

118 Burger, Robert S. How to Write So That People Can Under-
        stand You. Wayne, Pa.: Management Development In-
        stitute, 1969. 235p.
This book is for professionals who must write reports and wish to

improve their writing. It is composed of four parts: "The Lead,"
"The Background," "Elaboration of the Lead," "The 'Anything Else.'"
Within each part are subsections, within these are chapters (34 in
all), and within chapters are numbered subsections. The numbering
is not consecutive. Most chapters end in sentence exercises and an
answer key. The chapters cover words, sentences, and phrases.
Larger report concepts are not covered. A glossary lists grammatical terms. The author suggests that one can abandon the "confusing
advice" of past English teachers and follow his clear tips instead.
He refers to "schoolmarms" and says that in school if one communicated "clearly, simply, and concisely they thought you'd communicated
badly and discouraged you with a C or C- from doing it again."

119 Business Communication Case Book. Urbana, Ill.: American
          Business Communication Association, 1974. 126p.
This publication is a project of the 1973 Reports, Letters, and Cases
Committee of the ABCA. The committee revised cases already on
file at the ABCA and added new ones. "Part I, Revised Cases" includes "Letters" (five cases) and "Reports" (15). "Part II, New
Cases" includes "Letters" (five cases), and "Reports" (seven) and
"Mini-Communicative Problems" (five). Each case has the byline of
the instructor who created it. Examples include writing a product
letter to a fellow student and writing a memo giving a solution to a
problem of turnover in the steno pool. Cases include analyzing personnel problems, writing promotional material, reporting significant
factors, and recommending improvements. Some long cases have
notes for the instructor giving ways to make assignments on the case
or typical problems students encounter in trying to handle the problem. The depth and scope of the problems vary. Solutions are offered for some problems.

120 Business Communication Casebook Two. Urbana, Ill.: Ameri-
          can Business Communication Association, 1977. 78p.
This book, compiled similarly to Business Communication Casebook,
contains two parts: "Letters and Memoranda" and "Report Problems."
Cases in Part I vary from credit letter situations to developing a
questionnaire to sales letters. In Part II, the situations range from
job description of hotel employees to analyses of the medical-electronics
market and the feasibility of establishing a new restaurant.

121 Business Correspondence Library. 3 vols. Chicago: System,
          1911.

122 Butler, Ralph Star, and Burd, Henry A. Commercial Corres-
          pondence. New York: Appleton, 1919. Index. 531p.
Designed around a course developed at the Extension Division of the
University of Wisconsin, this book aims to teach students "the rules
of grammar and rhetoric" applied to business English and business-
letter form, organization, and strategy. The first two of the 24
chapters cover word choice and sentence/paragraph structure. Chapter 3 discusses letter format. The next four chapters cover general
letter-writing principles and letter types. The next two chapters deal
with complaint and adjustment letters. Four chapters on credit and

collection follow the chapters on job-application and "inspiration" letters (pep talks). The next nine chapters deal with sales letters. The final chapter covers form letters. The two appendixes are a grammar review and a punctuation guide.

123 Butler, Robert A. Handbook of Practical Writing. New York:
         Gregg/McGraw-Hill, 1978. Index. 234p.
Aimed at occupational students, and emphasizing basic English mechanics (grammar and punctuation) and paragraph construction, this text is devoted to general rhetorical questions of coherence, sentence building, and correct mechanics. There are two chapters in Part I directly bearing on business communication: "Business Letters and Memos" and "Getting a Job." Part II is entirely devoted to punctuation and basic grammar. Each chapter has exercises with answers.

124 Butterfield, William H. Bank Letters: How to Use Them in
         Public Relations. Danville, Ill.: Interstate, 1946. 68p.
According to the author, the reason letters are so important in public relations is that they are a one-to-one communication with the bank customer. Chapter 1, "Letters That Build Good Will for Banks," covers a few principles with ten sample letters. Chapter 2, "Selling Bank Service Through Personalized Letters," is a reprinted article by R. E. Doan advocating a personal style in bank letters. Chapter 3, "Let's Write Like Human Beings," points out the problems of writing in a stiff, impersonal style. Chapter 4, "Making Bank Patrons Feel Welcome," gives advice on winning new customers and regaining old ones. Chapter 5 contains sample letters covering public relations situations a banker is likely to encounter. The five appendixes serve as a handbook of forms of address and commonly misused words.

125 Butterfield, William H., ed. Better Customer Relations by
         Letter. New York: National Association of Credit Men,
         1945. 90p.
This collection of 11 articles is intended as a guide "for those engaged in handling the daily problems of the credit department." The articles were originally published in Credit and Financial Management. The book contains biographical sketches of the six authors represented in the collection. Selections include "Better Business Correspondence," by C. Chandler Parkhurst; "Courtesy Letters That Count," by the editor; and "Why a 'Routine' Letter?," by Wilbur K. McKee.

126 Butterfield, William H. Building Hotel Business by Letter.
         Stamford, Conn.: The Dahls, 1945. Index. 131p.
This book is primarily a compilation of effective letters from hotels. Seventy-seven hotel executives are listed as contributing material. The book is divided into three parts. Part I, "Letters to Patrons and Prospective Patrons," covers thank-you letters, solicitation letters, letters to local professional men, and letters extending courtesy on handling difficult situations. Chapter 7 covers the "Three C's of Effective Hotel Letters"--character, conciseness, cheerfulness. Part II, "Letters to Professional Associates," covers letters to hotel executives in other cities. Part III, "Letters to Employees," covers

a variety of employee situations.   Each chapter begins with a brief
introduction to the topic.

127   Butterfield, William H.   The Business Letter in Modern Form.
            enl. ed.   New York: Prentice-Hall, 1941.   Index.   Bib.
            302p.
This book is directed toward the business manager, the secretary,
and the student.   Chapters include "The Heading" (Chapter 4) and
"The Identification Line" (Chapter 10).   The final chapters treat
"Miscellaneous Structural Devices, " "The Envelope Address, " "Prop-
er Folding and Insertion in the Envelope, " "Stationery, " and "The
Letterhead. "   The appendixes cover abbreviations of state names,
"Correct Form in Writing to Army and Naval Officers, " etc.   A
short grammar/usage handbook is included.

128   Butterfield, William H.   Common Sense in Letter Writing:
            Seven Steps to Better Results by Mail.   Englewood Cliffs,
            N. J. :   Prentice-Hall, 1963.   Index.   238p.
The seven "Steps" in this book are chapter-length units titled, "Begin
By Getting All the Facts, " "Be Sure You Say What You Mean, " "Say
It--Don't Take Half a Day, " "Keynote Each Letter with Courtesy, "
"Focus Your Message on the Reader, " "Make It Sound Friendly and
Human, " and "Remember the TACT in ConTACT. "   There are checklists on the seven C's of letter writing (correct, clear, concise,
courteous, constructive, conversational, considerate); seven steps to
good business; ten pointers on good public relations by mail.   Several
poems and excerpts are highlighted in boxes.   "Letters That Pay Extra Dividends in Human Relations" is a section of model letters.   The
"Reference Section" contains advice on word choice, punctuation, usage, misspelled words, titles of address, and letter layout.

129   Butterfield, William H.   Credit Letters That Win Friends.
            Norman: University of Oklahoma Press, 1944.   Index.
            98p.
This book is designed to give a credit manager advice for "making
the most of his public-relations opportunities. "   The book analyzes
and illustrates the "functions of a 'top'notch' credit manager. "   In
addition to the many samples included, some of the chapters give a
problem with a poor solution, followed by an effective one.   The
seven sections of the book are: "How to Say 'Yes' Effectively, "
"How to 'Sell' the Credit Privilege, " "The Credit Man as a 'Trouble
Shooter, '" "Credit Letters Build Goodwill, " and "Smoothing Out Rough
Spots in Consumer Relations. "

130   Butterfield, William H.   Goodwill Letters That Build Business.
            New York: Prentice-Hall, 1940.   Index.   300p.
The first three chapters of this book deal with the theory of building
business through goodwill.   Chapters 4-7 demonstrate different kinds
of letter-writing situations, from adjustments to collections.   Chapter 8 provides a dozen "don'ts"--what to avoid, the psychological
"turn-offs, " and the linguistic "tactlessnesses. "   The book contains
a large number of sample letters from corporations and small businesses.

131   Butterfield, William H.   How to Use Letters in College Public
          Relations.   New York: Harper and Brothers, 1944.   In-
          dex.   182p.
This book is aimed at helping "college administrators ... prepare
for higher education on a grand scale ... [and] win staunch friends
for their institutions." Much of the book is composed of model letters and cases with evaluative commentary.   Chapter 1, "The Human
Touch Gets Results," discusses avoiding officialese, wordiness, and
using the "you attitude." The other chapters focus on groups that
administrators might write to: alumni, parents, prospective students,
faculty, and the business community.   An appendix is devoted to administrators in junior colleges.

132   Butterfield, William H.   How to Write Good Credit Letters.
          enlarged ed.   St. Louis: National Retail Credit Associa-
          tion, 1947.   Index.   116p.
This book deals with the public relations inherent in credit communications.   The 25 chapters are divided into two sections: "Letter-
Writing Fundamentals" and "Specific Types of Letters." Section III,
"Appendix," has samples of letter format and a list of misused expressions.   Chapter titles include "Negative Words Cripple Your Letters" and "Put Life and Friendliness into Your Letters."

133   Butterfield, William H.   Letters That Build Bank Business.
          Danville, Ill.:   Interstate, 1953.   Index.   100p.
This book is based on 18 articles that appeared in Banking, 1951-
1953.   The 12 chapters include such titles as "Good-Will Letters
Are a Sound Investment," "Make Your Holiday Letters Build Good
Will," and "Direct Mail at Work in a Progressive Bank." Each
chapter is primarily a set of sample letters introduced by brief comments on effectiveness.   Most chapters include short checklists, some
reprinted from other publications.   Also included is a subject index
to the sample letters.

134   Butterfield, William H.   Practical Business Letter Problems.
          New York: Prentice-Hall, 1938.   Bib.   112p.
This casebook-workbook contains 50 cases divided into 11 sections.
All the cases are on various letter types, e.g., acknowledgments,
sales, credit, collection, and adjustment.   The author includes a
general list of "Grading Standards" for the instructor.   The standards
cover layout, grammar and style, organization, and tone.   Many
problems include a sheet with appropriate letterhead so the students
can write their letters in the workbook.   There is space also for instructor comments after each problem.

135   Butterfield, William H.   Practical Problems in Business Cor-
          respondence.   New York: Prentice-Hall, 1942.

136   Butterfield, William. H.   Problems in Business Letter Writing.
          New York: Prentice-Hall, 1951.

137   Butterfield, William H.   Successful Collection Letters.   New
          York: McGraw-Hill, 1941.   Index.   250p.

This book suggests that appealing to the "humanness" in debtors is a sure way of getting them to pay. The book discusses the validity of the "you principle" in the first four of the nine chapters. The book also cites numerous examples of "poor" collection letters that insult or threaten the debtor. Chapter 7 gives "Fifty Tested Collection Letters" that illustrate the effectiveness of the strategies the book recommends. Chapter 9, "An Ounce of Prevention," deals with the wisdom of writing notes of appreciation to those customers who regularly pay their bills.

138 Butterfield, William H. Tested Credit and Collection Letters. Danville, Ill.: Interstate, 1950. 48p.
This collection of 100 model letters was first published by the National Retail Credit Association "as an aid to credit executives in building better consumer relations by mail." The ten chapters include "Granting Requests for Charge Accounts," "Expressing Thanks for First Credit Patronage," "Inviting Service Suggestions from Customers," and "Requesting Payment of Past-Due Accounts."

139 Butterfield, William H. 12 Ways to Write Better Letters. 2nd ed. Norman: University of Oklahoma Press, 1945. Index. Bib. 186p.
This book covers what the author describes as "the twelve most treacherous pitfalls of business letter writing." Each pitfall is covered in a separate chapter. Topics are correctness, naturalness, courtesy, simplicity, brevity, openings, completeness, the "you attitude," revision, tone, writing to one person at a time, and conclusions. Each chapter contains both good and bad specimen letters.

140 Calnan, James, and Barabas, Andras. Writing Medical Papers: A Practical Guide. London: Heinemann, 1973. Index. Bib. 121p.
This small book (5" x 6") is aimed at professionals who are writing papers. The seven chapters are divided into two parts. Part A, "The Authors," covers "The Junior Doctor," "The More Senior Doctor," and "The Consultant." Topics include examination papers, selecting journals, readability, style, applying for research funds, co-authors, publishing after the lecture, editorials, reviewing books. Part B, "The Craft," covers "Preparing to Write" (introduction, illustrations, abstract), "Writing the Manuscript," (revision, punctuation, clarity, jargon), and "The 10 Commandments," (a one-page list of reminders). The authors are on the staff of the Royal Postgraduate Medical School and Hammersmith Hospital, London. The staff of this institution produces "over 600 publications in reputable journals all over the world every year."

141 Campbell, Benjamin. Modern Business Punctuation. Jackson, Mich.: Business English, n.d.

142 Campbell, Benjamin J., and Vass, Bruce L. Brief Business English. Jackson, Mich.: Business English, n.d.

143 Campbell, Benjamin J., and Vass, Bruce L. Business Letters:

How to Write Them.   Jackson, Mich.: Business English,
n. d.

144   Campbell, Benjamin L. , and Vass, Bruce L.   Essentials of
        Business English.   Jackson, Mich.: Business English,
        n. d.

145   Campbell, Benjamin J. , and Vass, Bruce L.   National Business
        Speller.   Jackson, Mich.: Business English, 1921.   In-
        dex.   120p.
This work begins with general spelling principles and a pronunciation
guide.   Following are 104 lessons consisting of lists of words, divided into syllables, and their meanings.   The first 58 lessons cover
words grouped into classes, e. g. , "Words Ending in OR and ER. "
Beginning with Lesson 59, the stress is on words used in business.
Some lessons cover words connected with specialties, such as real
estate, insurance, law, automobiles, and hardware and lumber.   Lessons 97-100 cover cities and states.   The last four lessons cover
200 "demon" words.   The index is an alphabetical list of all the
words covered and the lesson in which they appear.

146   Campbell, H. D.   Campbell's Commercial Correspondence: The
        Art of Modern Business Letter Writing.   Los Angeles:
        Citizens Print Shop, 1918.

147   Campbell, John Morgan, and Farrar, G. L.   Effective Commu-
        nications for the Technical Man.   Tulsa, Okla.:   Petro-
        leum Publishers, 1972.   273p.
This book is geared to the professional technical writer in the petroleum industry.   The 14 chapters cover such topics as "Clearer,
More Effective Writing, " "Writing the Report, " "Technical Papers
for Publication, " "Mathematical Presentations, " "Verbal Communication, " "Error Analysis. "   The two appendixes are (1) a list of con-
ventional abbreviations, hyphen use, and special terms, and (2) a
list of standard letter symbols for Petroleum Reservoir Engineering,
Natural Gas Engineering, and Well Logging.   The examples throughout are relevant to the petroleum industry.   Some chapters end in
further references.   Other chapters end in writing exercises.   Sentence exercises are followed by the correct answers.   Chapters 11-
13 deal with correct use and presentation of statistical data.

148   Campbell, John S.   Improve Your Technical Communication.
        Los Angeles: GSE, 1976.   216p.
This book is composed of nine "lessons" (chapter-length sections)
first written as correspondence lessons for the Grantham School of
Engineering's home-study course.   The book takes the approach of
a self-study program, complete with tests and answer sections.
After the introductory lesson, the book includes two lessons on grammar and language.   The next five lessons deal with types of technical communication: reports, manuals, articles, proposals, and instructional materials.   The final lesson explains the editing process.

149   Candee, Alexander M.   Business Letter Writing.   New York:

Biddle, 1920.   347p.

This book is designed to present the "correct mental principles" that
design letters that "up-build for the good of the sender." Chapter 1
covers the importance and purposes of letter writing; Chapter 2 discusses diction and basic sentence grammar. Chapter 3, "The Principles of Thinking," presents the theory of business communication,
including logic and methods of development. Chapters 4-23 cover
types of letters (sales, collection, etc.) and writing strategy ("you
attitude," conciseness, goodwill, clarity). Although the book has no
pedagogical materials, it is the result of the author's courses in
commercial correspondence at the Extension Division of the University of Wisconsin.

150   Carr-Ruffino, Norma.   Writing Short Business Reports.   New
        York: Gregg/McGraw-Hill, 1980.

151   Cassell, R. J.   Art of Collecting; the Underlying Principles
        and Practices of Collecting, with Suggestions, Forms of
        Reports, Letters, etc. for the Collection Manager and the
        Business Man.   New York: Ronald, 1913.

152   Chappell, R. T., and Read, W. L.   Business Communications.
        3rd ed.   London: Macdonald and Evans, 1974.   Index.
        Bib.   214p.

This British textbook has 12 chapters on such topics as "The Spoken
Word," "The Written Word," "Business Correspondence," "Report
Writing," and "Summarizing Information." Chapter 11, "Aspects of
External Communications," covers telephone use, public relations,
company-meeting advertisements, and consumer and consulting coun-
cils. Chapter 12, "The Teaching of Communications," discusses
methods for the formal lecture, informal teaching, grading, and using
audiovisual aids. The first 11 chapters end in writing assignments.
Appendix I is a collection of sample examinations of the Royal Society of Arts in Communications and Report Writing of the Road Trans-
port Certificate. Appendix II is a bibliography.

153   Cherry, Charles L., and Murphy, George D.   Write Up the
        Ladder.   Pacific Palisades, Calif.:   Goodyear, 1976.

154   Chisholm, Cecil, ed.   Communication in Industry.   London:
        Business Publications, 1955.   Index.   Bib.   284p.

This book, primarily written for business executives in England, is
composed of 18 articles on a wide range of subjects, including a
company analysis of communication problems, writing speeches, and
using closed-circuit television. Of four sections, Section II deals
with written communication and includes "Survey of Media Available"
(house organs, annual reports, and personal letters), "Conveying
Your Meaning," (semantics and tone), and "Form, Style and Expression" (minutes, instructions, reports, and rules for "clear English").

155   Clapp, John Mantle.   Doing Business By Letter: A Complete
        Guide.   Vol. I.   New York: Ronald, 1935.   Index.   459p.

This book includes "600 illustrative letters" that reveal "expert handling" of the situations dealt with, each "chosen because it shows some essential idea involved in meeting the circumstances under which it was written." According to the author, the book includes "types of letters that never before have been dealt with in any work on business correspondence, such as the 'policy' letters of key executives, administrative letters dealing with questions of organization morale and discipline, and examples of the communications exchanged between organizations at each stage of extended negotiations." The 29 chapters are divided into seven parts. Part I, "Basic Essentials," covers writing to the reader, and style; Part II, "Language and Form of Letter," covers effective language and format; Part III-VI cover types of letters: sales, credit and collection, adjustment and complaint, purchasing and information, administrative and departmental.

156 Clapp, John Mantle. Doing Business by Letter. Vol. II. New
        York: Ronald, 1935. Index. 297p.
The second volume of this work "presents specimens of letters written in situations paralleling those illustrated in Volume I." The five divisions present model letters with a brief analysis under the following topics: "Specimens of Sales Letters," "Specimens of Credit and Collection Letters," "Specimens of Complaint and Adjustment Letters," "Specimens of Purchasing and Information Letters," and "Specimens of Administrative and Departmental Letters."

157 Clapp, John Mantle. Personal Letters in Business. New York:
        Ronald, 1935. Index. 337p.
According to the author, "These are letters written to business or official acquaintances, prompted by social or business obligations.... They are not business letters in the ordinary sense nor yet are they exactly personal letters. They are, in a way, business letters in the form of personal letters." The author further explains that this kind of letter requires "a technique which combines the accuracy, prudence, orderliness, and clearness of business letters with something of the informality and ease of personal letters." The 21 chapters in four parts cover "Letters of Courtesy," "Letters Relating to Requests," "Information Letters," "Telegrams, Resolutions, Forms of Address." There are very brief introductions to the divisions of model letters.

158 Clarke, Emerson. A Guide to Technical Literature Production.
        River Forest, Ill.: TW, 1961. 182p.
This book is written for technical writers who manage groups, estimate costs, recruit technical writers, write for military agencies, and use technical writing agencies. The book is divided into two parts. Part I, "The Elements of Production," describes "the organization, operation, and management of technical publication groups." Part II, "The Technical Writer," describes "special problems of recruiting, evaluating and upgrading the technical writer." The 17 chapters include "Engineer-Writer Liaison," "Recruiting the Technical Writer," and "Work-Area Layout and Environment." There are two appendixes: Appendix A, "Answers to Test Questions" of the earlier writer's aptitude test and Appendix B, "Artwork Sizing Chart." Following the appendixes is a ten-page supplement covering proposals and technical books.

159 Clarke, Emerson. How to Prepare Effective Engineering Pro-
         posals: A Workbook for the Proposal Writer. River
         Forest, Ill.: TW, 1962. 212p.
Lacking exercises, either fill-in or discussion, this book is not a
workbook in the traditional sense.   There are a couple of blank pages
for "Notes" after each chapter.   Part I covers organization, planning,
and front and back matter of the proposal.   Part II covers methods
for the efficient production of proposals, and Part III reviews the
previous topics in two lists: "Pre-mailing Check List" and "Proposal Evaluation."   The book is illustrated with cartoons, charts, and
tables from company proposals soliciting defense contracts.

160 Clarke, Emerson, and Root, Vernon. Your Future in Technical
         and Science Writing. New York: Richards Rosen, 1972.
         Index. 162p.
The first chapter of this career guide gives the definitions, rewards,
and contributions of technical and science writers.   Chapters 2-3
cover where and how the reader might work.   Chapter 4 describes
the primary skills needed to become a technical or science writer;
knowledge of science and technology, writing skill and secondary
skills, such as typing, diplomacy, and self-management.   Chapters
5-6 describe the fields of technical writing and science writing.
Chapter 7 describes the value of these fields for women.   Chapters
8-9 suggest ways to prepare for and get a job.   Chapter 10 lists
groups that can help the beginner get started, and Chapter 11 gives
general advice on getting started and why you might write.   Appendix A gives sources of education; Appendix B gives more about associations, addresses, and lists of professional technical societies.

161 Classen, H. George. Better Business English. New York:
         Arco, 1966. 108p.
This informal guide for executives has 12 chapters with such titles
as "Writing in Our Lines," "Demon Noun," "Missing Link," "Action,
Please!," and "Whom Are You Trying to Impress?"   Chapters include cartoons and numerous sentences taken from businesses.

162 Clausen, John. Mercantile Correspondence in English and Ger-
         man Languages. 22nd ed. 2 vols. Leipzig: Gloeckner,
         1919. Vol. I, 274p.; Vol. II, 272p.
This book is a reader in training people in business how to read
and write letters in German and English.   One volume is in English
and the other in German, covering identical material so that there
is exact translation of each letter.   The reader is advised to study
the letters for language proficiency.   The book contains 408 letters
drawn from several types of business: freight, insurance, banking,
and goods-trade.   The letters are those of inquiry, credit, offers,
orders, payment, exchange, and claim.   There are 44 pages of
commercial terms at the end.   Each letter is footnoted for difficult
words and phrases.   The title pages identify these volumes as the
22nd edition; however, the preface states, "This present volume is
the twenty-first edition ... in the course of the last thirty years."

163 Clifford, W. G. Building Your Business by Mail. Chicago:
         Business Research, 1914.

164 Cloke, Marjane, and Wallace, Robert. The Modern Business
        Letter Writer's Manual. Garden City, N. Y. : Double-
        day, 1969. 215p.
This book is a how-to-do-it guide for writing letters, complete with
models and plan sheets. The introductory sections give "ten commandments" and "things to think about" for writing letters. The
first four chapters cover planning, openings and closings, and clarity. Chapter 5 covers plain and simple word choice. Chapters 6-
13 cover types of letters. Chapter 14 discusses oral presentation.
Chapter 15 illustrates some unusual letters ("Yes, Virginia"). Chapter 16, "Tidbits of Information for That 'Gal Friday, '" gives letter
format.

165 Cobb, Charles. Practical Communication for Technical and
        Vocational Arts. Santa Monica, Calif. : Goodyear, 1978.
        Index. 204p.
This text-workbook with tear-out pages has 13 chapters covering such
topics as "The Basics of Communication, " "Work Orders and Esti-
mates, " "Other Necessary Forms and Records, " "How to Give Directions, " "How to Find References the Easy Way, " "Visualizing
Concepts and Data, " and "The Written Report. " All chapters end
in a "Learning Quiz, " which reviews the text and projects that require students to write a response, complete a form, or prepare
graphics materials. There are six appendixes covering grammar,
punctuation, and vocabulary. There are many examples of forms
for memos, work orders, job records, insurance claims, maintenance requests, etc.

166 Cochran, Wendell; Fenner, Peter; Hill, Mary, eds. Geowrit-
        ing: A Guide to Writing, Editing and Printing in Earth
        Sciences. Washington, D. C. : American Geological In-
        stitute, 1973. 80p.
The 20 sections of this book follow, roughly, the order of publication
--writing through editing to printing. In addition, to the standard
topics about writing, this book also covers such topics as writing
reviews for journals and book publishers, rules for names used in
the earth sciences, and how to choose a style guide. The book
gives numerous footnote and bibliography samples. It ends with a
detailed checklist for writing, which is referenced to the page where
the subject is covered.

167 Cody, S. English for Business Uses and Commercial Corres-
        pondence. Chicago: School of English, 1914.

168 Cody, S. How to Do Business by Letter. Chicago: School of
        English, 1908.

169 Cody, S. Success in Letter Writing. Chicago: McClurg, 1913.

170 Coffin, Royce A. The Communicator. New York: Barnes and
        Noble, 1975. 164p.
This book is a cartoon guide for success in business communication
at all levels. Each of the seven chapters begins with an introduction
summarizing the principles being stressed. Then follow cartoons

illustrating the precepts.   Topics include philosophies, fundamentals,
courtesy, telephone, writing, mannerisms, and techniques.

171   Cohen, M.   Planning, Preparing and Presenting the Technical
            Report. New York: Macmillan, 1968.

172   Colby, John, and Rice, Joseph A.   Writing to Express.   Min-
            neapolis: Burgess, 1977.   Index.   Bib.   134p.
This book of general advice is purposefully chatty: "Some who read
this book will find it less stuffy than they might have expected. " The
chapters are brief and informal, covering both business and technical
writing.   Chapter titles include "From the Top Brass, " "Like Mayon-
naise, " and "Or Would You Rather Be Raped?" The 24 chapters
are followed by an annotated bibliography that describes the way the
authors have used the books on the list.

173   Coleman, Peter, and Brambleby, Ken.   The Technologist as
            Writer: An Introduction to Technical Writing.   New York:
            McGraw-Hill, 1969.   Index.   Bib.   356p.
The authors of this textbook suggest ways for English teachers to
prepare themselves to teach technical writing.   The book has 11
chapters in three parts: "Selection of Data, " "Arrangement of Data, "
and "Presentation of Data. "   Each chapter has exercises for class
discussion and writing projects.   Topics include "Graphic Aids, "
"Specifications, " "Formats, " "Exposition-Description of Processes, "
"Tests of Evidence. "   There are two appendixes: a sample technical
report and standards for graphic presentation.   There are eight cases
of varied difficulty.   The authors state that omissions in the cases
are on purpose--the student is expected to exercise judgment and
develop approximations.

174   Collett, Merrill J. , et al.   Streamlining Personnel Communica-
            tions.   Chicago: Public Personnel Association, 1969.

175   Collier, Robert.   The Robert Collier Letter Book.   6th ed.
            New York: Prentice-Hall, 1950.   Index.   463p.
This book was originally published in 1931.   The author states that
"the greatest value of a book such as this [is that] it gives ... scores
of letters, on all manner of subjects, that have proved highly successful. "   The book contains hundreds of examples of sales letters.
Chapter 9, "The Six Essentials, " presents, in essence, Collier's
method for successful sales letters: (1) the opening that gets the
reader's interest; (2) the description or explanation of the product
or service; (3) the motive for the reader; (4) the proof or guarantee
to reassure the reader; (5) the "snapper, " or penalty, if the reader
does not act; (6) the close that makes it easy for the reader to act.

176   Comer, David B. , III, and Spillman, Ralph R.   Modern Tech-
            nical and Industrial Reports.   New York: Putnam's,
            1962.   Index.   Bib.   425p.
This text is divided into three parts.   Part I, "Principles: The
Theory of Report Writing, " discusses the process of communication,
gathering facts, and the rhetorical modes.   Part II, "Procedures:

The Techniques of Report Writing, " covers organization and components of the formal report, layout, effective writing style, abbreviations and specialized vocabulary, graphics, revision and proofreading, and classification of types of reports.   Part III, "Practice:
Specimens and Projects," is divided into four sections.   Section A
is a checklist for the report, samples of various kinds of reports,
and student report topics.   Section B contains illustrations of parts
of the report and various types of charts.   Section C gives standard
abbreviations for scientific and engineering terms.   Section D is a
selected bibliography.   Each chapter in Parts I and II is followed by
exercises.

177   Committee on Business Communication.   Modern Business Com-
          munication.   Ed. Charles B. Smith.   New York: Pit-
          man, 1963.   433p.
The Committee on Business Communication consists of 37 instructors.
The 20 chapters of this text cover such topics as business-
correspondence objectives, mechanics of a letter, organization, language, tone, types of letters, order forms, job applications, reports,
personal business letters, dictation, and form letters and duplicating.
Each chapter has discussion questions and writing exercises.   The
book contains no preface or identification of the Committee on Business Communication.

178   The Communication of Ideas.   rev. ed.   Montreal: Royal Bank
          of Canada, 1972.   141p.
This collection of 11 monthly newsletters on the subject of writing
includes "Let's Put Words to Work, " "The Discipline of Language, "
"Writing for All Occasions, " "Writing Letters, " "Writing an Article, "
and "Writing a Report. "   The Royal Bank of Canada devotes some of
its monthly newsletters to the subject of communication and offers
copies to teachers and executives.

179   Cooper, Bruce M.   Writing Technical Reports.   Baltimore:
          Penguin, 1964.   Index.   Bib.   188p.
This book is designed to help British scientists, engineers, and students write reports.   The seven chapters do not cover basic grammar, but they do cover the use of logic, modern technical writing
style, and methods of report development.   The book also discusses
"correctness" and the process a writer must master for good usage.
Chapter 7, on using illustrations, was written by C. Baker.   Following the chapters are a checklist of report-writing principles, an appendix with sources for library research, and a bibliography of grammar, usage, and technical writing.

180   Cooper, Frank E.   Writing in Law Practice.   Indianapolis:
          Bobbs-Merrill, 1963.   Index.   Bib.   527p.
Originally published in 1953 under the title of Effective Legal Writing, this book grew out of the author's legal-writing course at the
University of Michigan Law School.   Part I consists of ten chapters
on legal-writing style with examples, opinion and letter writing,
pleadings, briefwriting, contracts, statute drafting, and will drafting.   Part II is composed of three case studies, which the author

says can be used for classroom work.   No other pedagogical devices
are included.   Using quotations from respected jurists, the author
attacks the characteristics of poor style.   He also suggests the use
of the Flesch formula for clarifying legal prose.

181   Cooper, Joseph D.   How to Communicate Policies and Instruc-
           tions.   Washington, D. C.:   Bureau of National Affairs,
           1960.   Index.   348p.
This book covers "planning, writing, illustrating, and publishing written instructions ... internal administrative and operating instructions
of business organizations, consumer how-to-do-its, technical manuals, training manuals and related types of literature" in three parts.
Appendixes A-D cover advice for writing special forms: organizational manuals, salesman's manual, employee handbook, purchasing
policy manual.   Appendix E gives a "Planning and Production Check
List for a loose-leaf manual. "

182   Cornett, W. N.   French Commercial Correspondence and Tech-
           nicalities.   London: Hirschfeld Brothers, 1911.

183   Cornwell, Robert C. , and Manship, Darwin W.   Applied Busi-
           ness Communication.   Dubuque, Iowa: Brown, 1978.
           Index.   427p.
Each of the 15 chapters has a summary at the end and begins with
a list of principles the student will cover.   There are six parts:
"Interpersonal and Organizational Communication, " "The Purpose of
Human Interaction, " "The Foundation of Writing Skills, " "The Techniques of Business Writing, " "Oral Communication, " "Report Writing. "   The five appendixes cover grammar, proofreading marks, abbreviations, and a sample report.

184   Course in Business English.   12 vols.   New York: Business
           Training, 1916.

185   Cox, C. Robinson.   Criminal Justice: Improving Police Report
           Writing.   Danville, Ill.:   Interstate, 1977.   Index.   Bib.
           293p.
This book covers the format and varieties of police writing: reports,
letters, minutes, and other records.   There are 12 chapters in two
parts.   Part I covers notetaking, the "face" page, the continuation
page, writing, and editing.   Part II includes a review of basic grammar, spelling, punctuation, abbreviations, and usage.   Each chapter
has a summary and a test.   Answers are in Appendix B.   Illustrated
are forms from police departments across the country and samples
of headings, reports (Appendix A), and a typical city report-writing
manual (Appendix D).   Appendix C gives a list of commonly misspelled words.

186   Cox, Homer L.   How to Write a Business Letter.   rev. ed.
           New York: Bell, 1966.   Index.   125p.
This book was formerly titled How to Write a Letter.   The 11 chapters are given catchy titles: for example, "Are You Wasting Friday?" "Long Sentences Can Sell You Short, " "Good Conductors, "

"Sandwiching Your Message," and "The Uninvited Guest." Topics include the positive approach, wordiness, how to organize correspondence, the "you attitude," diction, organization, and sales letters. Chapter 4, "Are You Wasting Friday?" gives advice on how to use the services of a secretary efficiently and advises business executives, when opening their mail, to organize it into piles--"Important," "Not So Important"--and then answer the important letters.

187   Craddock, F. W. Dental Writing. 2nd ed. Bristol, England: John Wright and Sons for the University of Otago Press, 1968. Index. 95p.
This work is designed to help practitioners write scientific papers, articles, reviews, and letters. The 23 chapters cover such topics as jargon, conciseness, affectation, bibliographical references, typography, titles, abstracts, reviews, and evaluations of professional literature. All examples are related to dental science and practice.

188   Cramp, H. Letter Writing, Business and Social. Philadelphia: Winston, 1914.

189   Cresci, Martha W. Complete Book of Model Business Letters. West Nyack, N.Y.: Parker, 1976. Index. 298p.
The 15 chapters of this book cover letters of request, instruction, assistance, proposal, advice, acceptance and refusal, compliments and reprimands, credit, complaint, advertising, and sales. Personal letters are also included. One chapter is devoted to appearance and format, and a reference list gives special titles and forms of address. With each letter, the author provides guidelines and tips for writing.

190   Crissey, Forrest. Hand-Book of Modern Business Correspondence. Chicago: Thompson and Thomas, 1908. 352p.
The author states that this book is a guide for students, stenographers, correspondence clerks, all "office men," and managers. The 24 chapters discuss how to analyze a letter; types of letters; filing letters; the pecularities of manufacturing correspondence; and the problems of internal, departmental, agency, and branch communication. One chapter covers the credit systems in country stores.

191   Croft, Kenneth. Commercial Correspondence for Students of English as a Second Language. New York: McGraw-Hill, 1968.

192   Cross, Louise Montgomery. The Preparation of Medical Literature. New York: Lippincott, 1959. Index. 451p.
This book aims to provide techniques that will aid a medical practitioner in writing for journals or for book-length studies. The readers are doctors or allied scientists intending to publish the results of scientific clinical investigation. The 40 chapters are divided into seven sections covering the publishing process as it actually occurs. Topics are "Planning the Paper or Book," "Gathering Material for a Paper or Book," "Writing the Journal Paper or Book," "Style,"

"Styling" (covering mechanics), "Illustration," and "Editing." The book has short chapters and contains numerous headings.

193 Crouch, W. George. Bank Letters and Reports. New York:
        American Institute of Banking, 1961. Index. 491p.
This book has 21 chapters divided into four sections. Section I,
"Building the Bank Letter," covers such topics as the relationship
between reader and writer, interesting letters, dictation and trans-
cription, placement of information in letters, beginnings and endings,
word choice, sentence structure. Section II, "Types of Bank Letters," covers different types of letters. Section III, "The Kinds of
Writing Within Banks," covers memos, conference reports, and bank
reports. Section IV, "Helpful Reference Material," covers grammar,
punctuation, bibliography format, and the importance of English for
professional growth. Each chapter has writing exercises.

194 Crouch, W. George, and Zetler, Robert L. A Guide to Tech-
        nical Writing. 3rd ed. New York: Ronald, 1964. In-
        dex. Bib. 447p.
This textbook, designed for students in engineering, science, and
allied fields, covers in nine chapters business and technical letters,
the formal technical report, informal and memo reports, the technical and semitechnical article, illustrations, composition principles,
and speaking. Most chapters have both discussion questions and
writing exercises. The examples used reflect the fact that the authors expect the students to have familiarity with general technical
language. There are three appendixes: an index to English usage,
a selected bibliography, and a general reading list.

195 Cunningham, Donald H., ed. A Reading Approach to Profes-
        sional Police Writing. Springfield, Ill.: Thomas, 1972.
        Bib. 138p.
This anthology contains 15 readings either on police procedure or on
writing police material. The articles move from theoretical to practical. Chapter 3 is "The Fog Index," by Robert Gunning. Chapter
5 is "The Technique of Reviewing," by John E. Dewey. Chapter 12
is "Handcuffing a Prisoner," by Herbert P. Vallow. Chapter 15 is
a bibliography for police-report writing. The chapters are divided
into four parts: "Thinking and Writing," "Special Forms of Writing,"
"Special Techniques of Writing," and "Bibliography of Literature on
Police Reporting and Police Writing." Each part has a brief introduction. Each selection ends in questions for class discussion or
writing assignments and a vocabulary list. The book's premise is
that the selections will be studied as writing models.

196 Cunningham, Donald H., and Estrin, Herman A., eds. The
        Teaching of Technical Writing. Urbana, Ill.: National
        Council of Teachers of English, 1975. Bib. 221p.
This anthology is designed for English teachers who must teach technical writing for the first time. It is divided into eight parts with
24 articles, including definitions of technical writing, the use of
metaphor in technical writing, and general discussions of the differences between teaching technical writing and teaching freshman

English. There is a bibliography of journals in the field, a list of
bibliographies, and a list of recent articles about the teaching of
technical writing.   Each article is preceded by an abstract.

197 Dagher, Joseph P.   Technical Communication: A Practical
       Guide. Englewood Cliffs, N. J.: Prentice-Hall, 1978.
       Index. 322p.
The first two chapters in this textbook defines technical communication and explains the interpretation of technical information.   Chapter
3 covers speaking and listening on the job.   Chapters 4-7 discuss
methods of development.   Chapter 8 covers the use of illustration,
and Chapter 9 provides tips and exercises to improve reading.   Chapters 10-11 cover letters, and Chapter 12 covers research and organization.   Chapters 13-15 treat various types of reports, and Chapter
16 is a grammar review, organized much like a handbook.   The five
brief appendixes cover misspelled words, postal abbreviations, numbers, suggested clear phrases, and metric measurements.   The book
includes exercises, cases, and writing assignments.

198 Damerst, William A.   Clear Technical Reports.   New York:
       Harcourt Brace Jovanovich, 1972.   Index.   338p.
This text opens with a section outlining obstacles to clear communication, covering topics like "Unclear Reason for Writing, " "Lack of
Candor, " "Lack of Empathy. "   The section on technical writing covers memos, reports, proposals, articles.   There is a "Technical
Writer's Handbook, " covering spelling, grammar, punctuation, diction, and a glossary of grammatical terms.   Each chapter closes
with exercises.   An additional section, "Special Problems, " contains
class discussion questions and writing assignments.

199 Damerst, William A.   Resourceful Business Communication.
       New York: Harcourt, Brace and World, 1966.   Index.
       527p.
This textbook is divided into five parts.   Part I, "Understanding Re-
sourcefulness, " includes chapters on effective letter writing and effective dictation, style, and form.   Part II, "Investigating Routine
and Complex Problems, " includes chapters on informal company
writing, research, and the formal report.   Part III, "Writing Business Letters, " includes credit, collection, and claim letters.   Part
IV, "Getting the Job You Want, " covers applications and résumés.
Part V, "Editing the Final Draft, " is a handbook of grammar, punctuation, and usage.   The 20 chapters all end in exercises for class
and for writing assignments.

200 D'Aprix, Roger M.   How's That Again?   Homewood, Ill.: Dow
       Jones-Irwin, 1971.   160p.
This book, designed for the corporate employee, gives advice in using written and oral communication skills to win recognition for ideas
and talents.   The 16 chapters are divided into four sections covering
understanding the business organization and working within it, the
mechanical aspects of communication, tips for communicating with
an internal audience, and external communication.   The cartoon illustrations and chapters cover such topics as getting word to the

top, preparing and making presentations, readers and listeners, memos, reports, and promoting one's career.

201   The Dartnell Collection of 250 Tested Sales Letters.   Chicago:
           Dartnell, 1964.   unp.
This collection of sample sales letters is in a binder, with tabs marking the seven sections.   The letters were selected on the basis of "the percentage of returns and sales" and "the adaptability of the idea or plan of the letter to other lines of business."   The sections cover general sales letters, letters for obtaining and answering inquiries, letters to revive inactive accounts, sales-letter series, and novelty sales letters.   Section VII, "Miscellaneous Sales Letters," includes 27 tips for successful letters and lists of suggested openings and closings.   Each letter is shown on company letterhead.   A boxed inset explains the use and results of the letter and gives credit to the writer or contributor.

202   Davey, Patrick J.   Financial Manuals.   New York: Conference
           Board, 1971.   77p.
This work is a report by an independent, nonprofit business-research organization.   It is designed "to assist executives who are faced with the task of preparing a new financial manual or updating an existing one."   The sections of this report cover the purposes, preparation and maintenance, contents, and physical characteristics of financial manuals.   The two indexes include a composite listing of manual subjects (not cross-referenced to the body of this report) and a detailed table of contents of a typical manual.   The book gives writing advice primarily by showing models of narrative sections of manuals, focussing on typical content, arrangement, and format.

203   Davidson, Henry A.   Guide to Medical Writing.   New York:
           Ronald, 1957.   Index.   338p.
This book is for the practicing medical professional who wishes to publish or present a paper.   The 18 chapters include "Stalking the Idea: How to Start an Article," "Opening Paragraph: How to Hook Reader Interest," "Case Histories: Write, Don't Telegraph," "Cacoethes Scribendi: How to Write a Book."   The book deals primarily with research methods, revision, use of graphics, and use of figures. Chapter 18 deals with selecting the right journal and includes a list of journals divided by specialty.

204   Davis, A. L., ed.   Commercial Correspondence for Students
           of English as a Second Language.   Washington, D. C. :
           Educational Services, n. d.

205   Davis, Dale Stroble.   Elements of Engineering Reports.   New
           York: Chemical Publishing, 1963.   Index.   Bib.   200p.
This book is a textbook and a guide for the working engineer.   Chapter 1 covers the history of report writing and the need for training and defines the engineering report.   Chapter 2 distinguishes between the college report and the industrial report.   Chapters 3-6 cover the report, including introduction, body, and conclusion.   Chapters 7-8 cover graphics, and Chapter 9 reviews principles of clarity and

conciseness along with grammar, punctuation, and spelling.   Chapter
10 consists of sample reports marked for corrections.   Chapter 11
explains technical editing and the role of the technical editor.   The
four appendixes are "Symbols, Abbreviations, Numerals"; "Precision
of Measurements"; "Spelling, Proofreading"; and "Additional Exercises. "   The book includes questions for study.

206   Davis, J. F.   English Composition and Business Correspon-
            dence.   New York: Sir Isaac Pitman & Sons, n. d.

207   Davis, R. , and Lingham, C. H.   Business English and Corres-
            pondence.   Boston: Ginn, 1914.

208   Davis, Richard M.   Thesis Projects in Science and Engineer-
            ing: A Complete Guide from Problem Selection to Final
            Presentation.   New York: St. Martin's, 1980.   Index.
            253p.
This book is designed as a self-help guide for a student preparing
a Master's thesis.   In addition to discussing specific format, research, and writing requirements, the author offers tips for working
with an adviser and a committee.   The 16 chapters also include discussions of parts of the thesis, use of visuals, oral presentation,
and documentation.   Following the chapters is an appendix with sample thesis pages.

209   Davison, Ad-Man [Emil Bayard Davison].   The Master
            Letter Writer.   3rd ed.   New York: Harper and Broth-
            ers, 1930.   Index.   313p.
Originally published in 1920 by the author, this book is divided into
two parts.   Part I, "The Science of Successful Letter Writing, " explains the principles of good letter writing and discusses different
types of business letters.   Part II, "The Three Hundred Master
Business Letters, " consists of model letters written especially for
this book, not taken from business.   The models are examples of
the business letters discussed in Part I, which includes a section
on letter types.   The author's real name is Emil Bayard Davison.

210   Dawe, Jessamon.   Writing Business and Economics Papers,
            Theses and Dissertations.   Totowa, N. J. :   Littlefield,
            Adams, 1965.   Index.   192p.
The focus of this book is on the field of business and economics,
covering special data-gathering techniques, tabular and graphic illustration, and the formats of papers in business and economics.
Other chapters cover the research process, selecting and defining
the problem, building a research design, outlining, writing style,
and interpretation.   Each of the 11 chapters is followed by a substantial bibliography.

211   Dawe, Jessamon, and Lord, William Jackson, Jr.   Functional
            Business Communication.   2nd ed.   Englewood Cliffs,
            N. J. :   Prentice-Hall, 1974.   Index.   510p.
Despite a 20-percent reduction in length from the first edition, this
book of 13 chapters remains a comprehensive treatment of "problem

solving, involving the communicator and his audience in an intimate
relationship. " Chapter 1, "The Information System of Business, "
sets the stage by focusing on the role of communication in business
and in decision making.   The next four chapters comprise Part I,
"Behavioral Treatment of Human Relations Messages, " and cover
image building through clarity, goodwill letters, analysis of the communication situation, persuasion, and handling negative messages.
Part II, "Decision-Making Business Reports, " consists of six chapters on writing reports in the context of the decision-making process:
planning, gathering information, and convincing the reader.   Part
III, "Resource Material for Personal Use, " contains two chapters:
the job search and writing term papers and reports.   These two
chapters do not contain exercises; all other chapters have assignments, usually based on cases.

212   Dawson, Presley C.   Business Writing: A Situational Approach.
        Belmont, Calif.: Dickenson, 1969.   Index.   354p.
This textbook of 25 chapters is divided into four parts.   Part I, "Introduction, " discusses the background for the study of business communication and business style.   Part II, "Preparation for Business
Communication, " deals with letters, tone, clarity, and organization.
All chapters in Parts I and II end in a summary, discussion questions, and writing assignments.   Part III, "Situations for Business
Writing, " begins with a checklist for letter writing, cross-referenced
to previous chapters.   The 15 chapters that follow cover types of
letters, such as application, inquiries, complaints, sales, and memos
and reports.   All chapters end in writing assignments.   Part IV,
"A Critical Approach and Review, " consists of one chapter giving a
four-page outline of an individual critique of a letter and brief men-
tion of a group critique.   Appendix I, "Selected Principles of Composition for Business Communication, " discusses general rules "most
likely to be broken by the average student, " e. g. , theme, paragraph,
sentence.   Appendix II, "An Interrelated Writing Project, " is a case
project with 14 assignments asking for most of the letter types covered in the book.

213   Day, Robert A.   How to Write and Publish a Scientific Paper.
        Philadelphia: ISI, 1979.   Index.   Bib.   160p.
According to the author, "the purpose of this book is to help scientists and students of the sciences in all disciplines (but with emphasis on biology) to prepare manuscripts that will have a high probabil-
ity of being accepted for publication and of being completely under-
stood when they are published. " The 26 chapters begin with discussions of "What Is a Scientific Paper?" and how to write various
parts of the paper, e. g. , title, abstract, introduction, materials
and methods, results, and discussion.   Following chapters cover
citing sources, using tables, preparing illustrations, and typing the
manuscript.   Chapters 15-22 deal with steps of the publishing process: submitting the manuscript, dealing with editors and reviewers,
handling reprints, and reviewing the articles of others.   Chapter 22
is concerned with ethics, rights, and permissions.   Chapter 23 covers general problems of grammar and structure, and Chapter 24
discusses jargon.   The last two chapters provide advice on using

abbreviations and a short commentary by the author. The six appendixes provide guidelines for abbreviations and symbols, words and expressions to avoid, and common errors in style and spelling.

214 Day, Stacy B. , ed. Communication of Scientific Information. Basel, Switzerland: Karger, 1975. 240p. This collection of 23 essays has been prepared to introduce the student, layperson, or professional scientist to a number of fields that "contribute to communication science. " Articles cover writing, information technologies, communication theory, and related subjects. One-third of the articles are devoted to medical and biomedical communication. Articles include "Medical Writing: Discrepancies Between Theory and Practice, " by W. B. Bean; "Science Literacy and the Public Understanding of Science, " by B. S. P. Shen; and "Idiolect, " by J. M. Dorsey.

215 DeBakey, Lois. The Scientific Journal: Editorial Policies and Practices. St. Louis: Mosby, 1976. Index. 129p. This style guide for those preparing an article for a scientific journal is divided into 31 chapters compiled by members of a group formed to survey editing practices in scholarly journals. Most of the members edit biological or medical journals. The chapters are divided among major categories: "Editorial Policies and Editorial Practices. " Under "Editorial Policies, " chapters are further divided into "Reviewing" and "Special Types of Manuscripts. " Topics include multiple publication, advertising, editorials, and book reviews. The category "Editorial Practices" contains further sections on "References" and "Format, " with chapters covering the journal cover, indexes, reference form, and binding. The three appendixes provide an organizational chart for a scientific-journal office, a manuscript flow chart, and the rules and checklist and form for preparing an abstract for the American Federation for Clinical Research.

216 DeCaprio, Annie. A Modern Approach to Business English. Indianapolis: Bobbs-Merrill, 1973. 219p. This text-workbook with tear-out pages is aimed at students whose background may be deficient in grammar, spelling, or pronunciation. Topics covered in the 24 units include agreement of subject and predicate, possessive nouns, adverbs, pronouns, commas, and apostrophes.

217 DeCaprio, Annie. A Modern Approach to Business Spelling. Indianapolis: Bobbs-Merrill, 1974. 224p. This text-workbook is aimed at students deficient in spelling, grammar, pronunciation, and English usage. Contents include dictionary entries, dictionary respellings, roots, dropping the final "e, " prefixes, compounds and hyphenation, and plurals.

218 Dederich, Robert M. Communication of Technical Information. New York: Chemonomics, 1952. Bib. 116p. This book was originally distributed only to clients and special groups that requested it of Chemonomics, an affiliate of R. S. Aries and Associates, Consulting Engineers and Economists. Publication was

prompted by the inability of technical personnel to communicate effectively or present information "when and where it is needed, and in the form it should take." The book is a guide to those who must write reports horizontally or vertically within the organization and address essential questions of objective, reader, and purpose. The seven chapters cover the nature of communication, written and oral reports.

219 Deffendall, Prentice Hoover. Actual Business English. rev.
        ed. 1923; rpt. New York: Macmillan, 1928. Index.
        Bib. 224p.
This text, first published in 1922, appears to be for high school commercial students. The word "actual" is used in the title because the author asserts that every sentence was taken from real businesses. The entire book is devoted to grammar and general rhetoric from nouns to the whole composition. Although examples are taken from business letters and the "you viewpoint" is mentioned in the discussion of paragraphs, very little space is given to letter-writing strategy. The exercises at the end of the book are usage drills.

220 deMare, George. Communicating at the Top. New York:
        Wiley, 1979. Index. 270p.
A revision of a work published in 1967 under the title Communicating for Leadership, this book aims to teach executives what they "need to know about communicating to run an organization." The 20 chapters of this book are divided into six parts. Part I, "Style," gives general advice on writing and contradicts what the author calls "myths" of writing. For example, he suggests that the idea that "a good writer knows how to use words" is not necessarily true. As proof, he cites Theodore Dreiser as a writer with "a tin ear for words" and William Faulkner as "an atrocious writer if measured by his tortured language." Throughout the book, literary figures and thinkers are frequently quoted. Part II, "The Mysterious World Below Formal Channels," deals with choosing forms of communication (letters, memos, formal reports, etc.). Part III, "Reaching Out," treats the various publics, and Part IV, "Putting It Together," discusses communication programs for small, medium, and large organizations. Part V, "Ruling the Minds of Men," examines the "art of thought" and ponders the future of communication. A concluding statement, "Envoi," ends by asserting, "If we are to reach other minds, we must be the ruler of our own."

221 DeMaris, R. E., ed. Readings in Science and Technology: An
        Approach to Technical Exposition. Columbus, Ohio: Merrill, 1966. 158p.
This collection of 18 readings is divided into five parts: "Some Perspectives on Writing and Thinking," "The Literary Approach to Science and Technology," "The Technical Approach: Definition," "The Technical Approach: Description of Mechanisms and Processes," and "The Technical Approach: Analysis and Interpretation." Authors represented include Rachel Carson, Thomas Henry Huxley, Lewis Mumford, Charles Darwin, Morris Freedman, and Robert Ardrey. Each selection is followed by discussion questions and, in most cases, practice exercises for writing.

222 Dermer, Joseph.   How to Write Successful Foundation Presen-
              tations.   rev. ed.   New York: Public Service Materials
              Center, 1974.   80p.
The Public Service Materials Center, established in 1967 "to meet
a need for useful and informative materials relating to fund raising, "
commissioned this book for those who work at fund raising from
foundations.   The book of six chapters covers three major steps:
appointment letter, proposal for grant, and letter of renewal.   One
chapter each is devoted to these three steps.   Actual letters are
used with identifications removed.   Commentaries following the cases
and sample letters and proposals are geared to letting the reader
know how to approach foundations and how to make persuasive appeals.   Two chapters deal with (1) the appeals for various projects,
and (2) the writing of a proposal for a building.   The last chapter,
"Conclusion, " summarizes the advice given both on writing and on
the persuasive techniques needed for appeal to foundations.

223 Devlin, Frank J.   Business Communication.   Homewood, Ill. :
              Irwin, 1968.   Index.   Bib.   705p.
This text has 18 chapters in five parts: "Business Communication
Principles, " "Planning and Organization, " "Business Letters, " "Business Reports, " "Supplements. "   Part V includes language review,
stenographic aids, and a glossary.   Each chapter ends in discussion
questions, writing exercises, and a summary.   Chapter titles include
"Encoding the Message, " "Dictating a Letter, " "Creating Sales by
Letter, " and "Streamlined Reports for Everyday Uses. "   The author
stresses the idea that business writing encompasses both communication and good business practice.

224 Dienstein, William.   How to Write a Narrative Investigation
              Report.   Springfield, Ill. : Thomas, 1964.   Index.   115p.
This book has five chapters presenting the essentials for police-
investigation reports, including types of information to be used,
principles of good report writing, and format of a report.   Four
appendixes give samples of a traffic accident report, a burglary report, a malicious mischief report, and a juvenile report.   Also included is a summary of all suggested guidelines.

225 Dirckx, John H.   Dx + Rx--A Physician's Guide to Medical
              Writing.   Boston: Hall, Medical Publications Division,
              1977.   Index.   Bib.   238p.
The 12 chapters of this guide are divided into three sections.   Part
I, "The Anatomy of Language, " reviews modern linguistics and
sources of medicine's technical language.   Part II is "Sentences and
Paragraphs: The Physiology of Language. "   The approach to gram-
mar is technical, with illustrations drawn from medical sources.
Also covered in this section are clarity, readability, accuracy, and
organization.   Part III, "Blunders and Blather: The Pathology of
Language, " covers word choice, syntax, and style.

226 Dirckx, John H.   The Language of Medicine.   Hagerstown, Md. :
              Harper and Row, 1976.

227 Director's and Officer's Complete Letter Book.   Englewood

Cliffs, N. J.: Prentice-Hall, 1965. 256p.
This model-letter book gives sample letters on the right pages with
general rules and alternate phrases on the left pages. The scope is
broad; the book covers nearly every writing situation. The 11 chapters cover letters for special situations, letters to stockholders,
personal letters, customer goodwill letters, order letters, sales let-
ters, credit and collection letters, letters involving community activities, and letters getting employee cooperation. Chapter 10 deals
with appearances, and Chapter 11 provides forms of address for officials and persons of rank.

228 Dodds, Robert H. Writing for Technical and Business Magazines. New York: Wiley, 1969. Index. 194p.
This book, written for the amateur, treats in detail the publishing
sequence: choosing the subject, determining the audience, planning
and writing the article, illustrating, preparing the manuscript, and
setting it. There is useful information on editorial practices. The
three appendixes cover style and copyright practice and include a
list of technical and business magazines that publish contributed articles.

229 Dolch, Edward William. Manual of Business Letter Writing.
New York: Ronald, 1922. 361p.
According to the author, "this book is intended to help business men
get the greatest possible results from their correspondence." To
make the book "useful for quick reference," the author has used the
format of an expanded outline with every paragraph a new section of
the outline. Written for both "instructors and correspondence supervisors," the book provides letters for correction and letter problems
in two of the five appendixes (the others contain hints for stenographers and punctuation and grammar/usage sections). The ten chapters
contain advice on salesmanship in letters, letter style, order and inquiry letters, claims and adjustments, credit and collection letters,
sales letters, job-search letters, memos and form letters, and format of letters. There is an index for the sample letters.

230 Donnelly, Austin. Communications in Modern Business. Sidney, Australia: Butterworths, 1960. Index. 280p.
Although the author says that this book is a text, there are no exercises, and it appears to be geared to the person on the job. The
30 chapters are divided into five sections. Chapters cover such
topics as "Business Writing Manuals," "Holding the Reader's Interest," "Sales Writing," and "Development of Business Writing." Appendix I is "Some Do's and Don'ts," a list from the manual of New
York Life Insurance Company. Appendix II is "The Simplified Letter
Form," which shows block style with no salutation or close. Appendix III is "Education in Business Writing in the U.S.A.," which contains two articles on business writing from the ABCA Bulletin: one
on the program at University of Texas and one on a survey of writing
courses. Appendix IV is "Writing for Publication," which is an article the author wrote for Australian Accountant.

231 Donnelly, Austin. How to Persuade People Through Successful

Communication and Negotiation.   Sydney, Australia:
Rydge, 1977.

232 Donovan, Peter.   Basic English for Science.   Oxford, England:
         Oxford University Press, 1978.   153p.
The 11 units of this guide for nonnative speakers of English in the
sciences are organized "so that the student proceeds from simple
tasks, such as expressing values and formulae in English, to the
descriptions of complete experiments, explanations, and accounts of
processes." Exercises include filling in sentences, correction, and
short writing assignments.   All examples are taken from basic mathematics and physics.

233 Doris, Lillian.   Modern Corporate Reports to Stockholders,
         Employees, and the Public.   New York: Prentice-Hall,
         1948.   Index.   309p.
A comprehensive, illustrated study, this book is aimed at those who
prepare, study, or evaluate annual reports.   The chapters not only
cover the purpose and scope of annual reports but also give advice
on parts of annual reports: the narrative section, financial section,
and letter to stockholders.   The emphasis throughout is on making
each part readable and interesting.   Other topics include building
goodwill among employees, stockholders, management, and the public.   The final chapter shows "attractive pages from annual reports."

234 Douglas, George H. , ed.   The Teaching of Business Communi-
         cation.   Champaign, Ill.:   American Business Communi-
         cation Association, 1978.   238p.
This collection of articles from The ABCA Bulletin (1972-1977) is
aimed primarily at people entering the field of business communica-
tion.   However, new teachers of technical writing will also be inter-
ested in many of the articles, such as "What of the Inexperienced
Instructor," by L. W. Denton; "The Basic Technical and Business
Writing Course at Georgia Tech," by Karl M. Murphy; and "Teaching
Writing in a College of Engineering," by Thomas W. Sawyer.   The
40 articles are divided into five parts.   Part I, "The Beginning
Teacher," covers approaching the classroom for the first time.
Part II, "Course Content and Curricula," includes course outlines
as well as articles about content.   Part III, "Teaching Methods and
Techniques," provides unique ideas for presenting material in the
classroom.   Part IV, "Grading Practices," focuses on methods for
grading and evaluation assignments.   Part V, "Teaching Aids," includes the use of media as well as "How to Write Problems," by
Francis W. Weeks.

235 Douglass, Paul.   Communication Through Reports.   Englewood
         Cliffs, N. J.:   Prentice-Hall, 1957.   Index.   410p.
This book was written for students and professionals, although inclusion of a syllabus suggests it came from the classroom.   The 26
chapters are divided into three parts: Part I discusses the "principles and techniques employed in the development of a report" (research, readability, clarity, charts); Part II is devoted to language
fundamentals (words, sentences, paragraphs, and punctuation); Part

III covers the types of reports (research, information, oral, staff study, legislation, public information). The author encourages and shows readers how to build their own handbooks. Writing and editing exercises are included at the end of each chapter. Appendix I covers "The Reporting of Accountants," and Appendix II gives a "Course Syllabus" geared to this textbook.

236 Dover, C. J. Effective Communication in Company Publica-
          tions. Washington, D. C. : Bureau of National Affairs,
          1959. Index. Bib. 367p.
This spiral-bound book contains 17 chapters divided into four parts. Topics include the role of communication as a management tool, controversial communication within organizations, the editorial process for large and small companies, how to evaluate publications, and the role of the editor. The three appendixes list the editors involved in the publications used as examples, a basic reference library for house-organ editors, and a supplemental bibliography. Many samples and case histories of company publications are included.

237 Dow, Roger W. Business English. New York: Wiley, 1979.
          Index. 451p.
"This book provides a program of study in English grammar and punctuation for a basic course in business English at the post-high school level." The textbook, which can also be used for self-study work, has 44 units divided into six sections. Each unit ends with drills. Section I covers using the dictionary, word division, parts of speech, and diagraming. Sections II-III cover punctuation, capitals, and figures. Sections IV-VI cover grammar. Section VI also reviews the earlier units.

238 Drach, Harvey E. American Business Writing. New York:
          American Book Company, 1959. Index. 495p.
The 25 chapters in this text are divided into two parts. Part I, "The Substance," covers the form and strategy for writing letters and reports. The 21 chapters in this section cover such topics as form letters and form paragraphs, analyzing your product, and interdepartmental communication. Part II, "The Medium," covers grammar, punctuation, spelling, and idiom. An appendix contains usage items and a series of questions with answers from the author's classes. Exercises at the end of chapters contain numerous cases.

239 Draughon, Clyde O. Practical Bank Letter Writing. Boston:
          Bankers Publishing, 1971. Bib. 217p.
This book of 16 chapters begins by discussing planning, style, appearance, and vocabulary. The middle chapters cover such topics as completeness, letter length, the "you attitude," and sincerity. The final chapters cover difficult letters: congratulation, condolence, and recommendation. Chapter 16 contains sample letters for various banking contexts.

240 Duddy, Edward A. , and Freeman, Martin J. Written Commu-
          nication in Business. New York: American Book Com-
          pany, 1936. Index. 527p.

This text had two objectives: to acquaint the business student "with the function of communication in business management," and to teach the "technique of writing communication forms." Part I, "The System of Communication," covers the theory and general practice of communication in business; Part II, "The Technique of Communication Forms," covers types of letters and reports, including the telegram and cablegram. Part III, "The Communication Problems of the Business Unit," covers special forms and principles of various departments in a business, such as purchasing and public relations. Following the 28 chapters are three appendixes: sample articles, the mechanics of report writing, and a sample report.

241 Durrenberger, Robert W. Geographical Research and Writing.
       New York: Crowell, 1971.

242 Dwyer, Ion E. The Business Letter. Boston: Houghton Miff-
       lin, 1914. Index. 177p.
This textbook advocates "a style that is free from meaningless formality." It is divided into five parts, subdivided into lessons. After a brief three-page introduction, Part I covers in six lessons the physical appearance and format of the business letter. Part II covers strategies for types of letters, such as orders, credit letters, form letters, and sales letters (nine lessons). Part III covers postal requirements and telegrams. Part IV is a problem-solving situation in which various letters are illustrated and the student responds with a letter. Part V is a seven-page section on filing letters. The Appendix contains a glossary of business terms, business abbreviations and symbols, and postal information.

243 Dyer, Frederick Charles. Executive's Guide to Effective Speak-
       ing and Writing. Englewood Cliffs, N. J.: Prentice-Hall,
       1962. Index. Bib. 240p.
The 22 chapters here are divided into two parts. Part I, "Effective Speaking for Executives," covers primarily the public speech in ten chapters. Part II, "Effective Writing for Executives," discusses the principles of writing letters, reports, and articles in 11 chapters. Among the subjects are managing the writing of others and the use of readability formulas. The two appendixes list materials and organizations for further reference.

244 Earle, S. C. Theory and Practice of Technical Writing. New
       York: Macmillan, 1911.

245 Easlick, Kenneth A., et al., eds. Communicating in Dentistry.
       Springfield, Ill.: Thomas, 1974. Index. 228p.
This spiral-bound manual was prepared by faculty members in the Schools of Dentistry and Public Health at the University of Michigan. It is intended as a guide for dentistry students preparing their scientific reports. Topics covered include sources and evaluation of the information, preparation of manuscripts, oral reports, research proposals, use of the dental library, bibliographic references, grammar, and recommended terminology for biological explanations.

246 Eckersley, Charles E. , and Kaufman, W. A. English and
         American Business Letters. London: Longman, 1974.

247 Effective Communication. Harvard Business Review Reprint
         Series No. 21073. Boston: Harvard University Press,
         1974.

248 Effective Communication for Engineers. New York: McGraw-
         Hill, 1975. Index. 216p.
This book is a collection of articles originally published in Chemical
Engineering. The articles give pragmatic advice for the engineer on
the job regarding writing reports, preparing reports for publication
and giving oral presentations. Topics include "Producing Visually
Effective Reports, " "Presenting Ideas to Groups, " "Working with a
Secretary, " "Getting Your Material into Print, " and "Ten Common
Weaknesses in Engineering Reports. "

249 Ehrlich, Eugene H. , and Murphy, Daniel. The Art of Techni-
         cal Writing: A Manual for Scientists, Engineers, and
         Students. New York: Crowell, 1964. Index. 182p.
This book is particularly for those who hated English 101 and vowed,
"When I get my grade in this course, I will never write another
page. " The first part of the book, "Technical and Scientific Writing, " covers in six chapters the types of documents the engineer
and scientist must write: abstract, proposal, letters and short reports, memos, and journal articles. The book also stresses responsibility in providing a good environment for effective writing by employees. The second half of the book is a "Handbook of Style and
Usage" including a glossary of usage and grammatical terms and a
chapter on spelling.

250 Eisenberg, Anne. Reading Technical Books. Englewood Cliffs,
         N. J. : Prentice-Hall, 1978. 241p.
Designed primarily for students in community and junior colleges and
beginning science students, this book could be used in conjunction
with a technical writing textbook. It aims to help students read with
comprehension and grasp the organization of technical writing. There
are three parts: "Basics of Technical Reading, " "Using a Technical
Book, " and "On Your Own" (a study-skills section). Samples come
from physics, data processing, and automotive, electronic, and mechanical engineering books.

251 Elfenbein, Julian. Handbook of Business Form Letters and
         Forms. New York: Simon and Schuster, 1972. 314p.
This book is a guide for secretaries and clerks who are expected to
handle company correspondence on their own. The book provides a
number of sample form letters, memos, proposals, bids, and résumés. The 22 chapters cover types of letters and memos, giving
samples and tips.

252 Ellenbogen, Abraham. Effective Business Correspondence.
         1963; rpt. New York: Macmillan, 1975. Bib. 128p.
Originally published as The Collier Quick and Easy Guide to Business

Letter Writing, this book contains 18 short chapters in a large format (8" x 10"). In addition to standard letter types, a chapter is
devoted to etiquette and social writing in business (sympathy, illness,
birth). Sections are included also on reproduction and duplication of
letters, telegraph services, and mailing. The four appendixes give
brief advice on punctuation, spelling, abbreviations, and filing.

253 Emberger, Meta R. , and Hall, Marian R. Scientific Writing.
          Ed. W. Earl Britton. New York: Harcourt, Brace,
          1955. Index. Bib. 468p.
The 15 chapters of this text cover three major topics: research and
organization, composition, and types of papers. The first six chapters cover the scientific method and its relationship to the writing
process, gathering data, and organization. Chapters 7-9 cover writing techniques. Chapters 10-15 focus on formats: research papers,
reports, abstracts, case histories, book reviews. Each chapter begins with a topic outline and ends in study questions. The two appendixes provide a suggested reading list in the theory, practice,
and types of scientific writing, and examples of letters.

254 Erskine, Frank M. Modern Business Correspondence. Indian-
          apolis: Bobbs-Merrill, 1920. Index. 176p.
This high school text consists of a series of "lessons." Each lesson
is devoted to a problem in business-letter writing and is followed by
student writing exercises. The 19 chapters include the parts of the
letter; the envelope; paragraphs; letters of credit, collection, application, remittance, introduction, and recommendation; telegrams;
filing; and advertising. Included are lessons on punctuation and usage. The appendix lists abbreviations for states and commercial
terms. The sample letters in the book represent the correspondence
of one business firm "covering a period of years. "

255 Estrin, Herman A. , ed. Technical and Professional Writing:
          A Practical Anthology. New York: Harcourt, Brace
          and World, 1963. Index. Bib. 317p.
This collection of 44 articles is divided into five categories. "Introduction" contains two articles on the importance of good technical
writing. "Planning--Procedure and Structure" covers style and
clarity. "Assembling Prose Components with Precision" covers the
mechanics of sentence structure and conciseness; "Gearing the Message to the Audience" and "Conclusion: Impact and Perspective" cover a large view of the past and present importance of clear technical
writing.

256 Evans, Gordon H. Managerial Job Descriptions in Manufactur-
          ing. AMA Research Study 65. New York: American
          Management Association, 1964. Index. 365p.
This study is designed to help manufacturing companies use and develop effective job descriptions for both management and hourly personnel. There are two parts. Part I (62 pages) provides background for using and creating job descriptions: their purpose, the
writing process, data collection, format, content, and flexibility.
Part II is composed of sample job descriptions at every level (e. g. ,

foremen, administrators, engineers). A final section of Part II is
an unabridged version of a company organization guide. Included are
an index of job descriptions and an index of company names used in
the study.

257 Ewer, J. R. , and Latorre, G. A Course in Basic Scientific
       English. London: Longman, 1969. 199p.
This book is designed primarily for non-native speakers of English
who have been studying English for two years. It uses an almost
programmed approach: brief explanations, examples, exercises.
The few narrative explanations of principles of writing are for the
teacher or for reading practice. The first 84 pages are divided into
12 units covering various linguistic principles, using repetition and
drill. Following are 18 readings--from 270 words to 990 words--
on various scientific subjects. Also included are four appendixes
and a 48-page basic dictionary. The appendixes cover prefixes and
suffixes, irregular verbs, abbreviations and symbols, and "Anglo-
American Weights and Measures with Metric Equivalents." Also included is an index of grammatical structures and terms. The authors
are an English professor and an engineering professor at the University of Chile.

258 Ewing, David W. Writing for Results in Business, Government,
       and the Professions. New York: Wiley, 1974. Index.
       Bib. 466p.
This book covers such topics as business reports, memos, letters,
proposals, summaries, forecasts, descriptions, and analyses. One
discussion centers on whether a communication needs to be written
at all, and there is lengthy discussion of the need to analyze the
reader's bias and his or her educational and psychological preparation for the communication. There are numerous checklists covering
such things as six common misbeginnings, nine rules of persuasion,
and 14 patterns to stress analysis and evaluation.

259 The Executive's Complete Portfolio of Letters. Waterford,
       Conn.: Bureau of Business Practice, 1967.

260 Eytinge, Louis Victor. Writing Business Letters Which Get
       the Business. The Pocket Book Series. Chicago: Office
       Applicance Company, 1914. 38p.
This book consists of six articles originally written for Office Appli-
cances, a magazine of office equipment. Chapter 1, "Get Into the
Envelope and Seal the Flap," advises naturalness and character in
letters. Chapter 2, "Get Under the Prospect's Hide," advises on
direct-mail sales ideas. Chapter 3, "Get a Persuasive Perspective
for Your Prospect," stresses the "you attitude." Chapter 4, "Get
a Good Grip on Your Prospect," covers persuasion. Chapter 5,
"Get the Dotted Line Signed and Get Away," gives effective closings
for sales letters. Chapter 6, "Get Good Associates," urges good
stationery, format, typing, etc. The author was a life-termer in
the Arizona Penitentiary for a "capital crime" after an earlier career of forgery. He is presented as a model example of rehabilita-
tion and says, "Three things have actuated me in the writing of this

series: First, --the fine fellowship that has been given me by the
editors and their friends. Second, --the love of good letters and my
desire to see letters more efficient. And third--the most important--
the desire to awaken you businessmen, you taxpayers, to sober
thought on one of the world's greatest problems--the prison and the
prisoner. "

261 Fahner, Hall, and Morris, E. Miller. Sales Manager's Model
       Letter Desk Book. West Nyack, N.Y.: Parker, 1977.

262 Famularo, Joseph J. Organization Planning Manual. New
       York: American Management Association, 1971.

263 Farmiloe, Dorothy. Creative Communication for Career Stu-
       dents. Toronto: Holt, Rinehart and Winston, 1974.

264 Fear, David E. Technical Communication. Glenview, Ill. :
       Scott, Foresman, 1977. Index. 436p.
This book has 18 chapters divided into three parts. Part I covers
in six chapters the basics of planning, organizing writing, revision,
graphics, and argumentation. Part II applies these basics to correspondence, reports, forms, memos, group and individual speaking,
and informal communication (telephone, shoptalk, nonverbal and listening). Part III serves as a handbook of grammar, mechanics,
usage, and research papers. Each chapter begins with a summary
and ends with exercises and planning worksheets. Military numbering is used.

265 Fear, David E. Technical Writing. 2nd ed. New York:
       Random House, 1978. Index. 257p.
The seven chapters in this textbook cover such topics as basic writing techniques, modes of technical writing, illustrations, letters,
informal/formal reports, oral reports, and research. Exercises
include writing assignments and sentence exercises. Appendixes
cover gathering information and documenting sources. Military
numbering is used.

266 Fielden, Frederick Joshua. A Guide to Precis Writing. Lon-
       don: University Tutorial, 1952.

267 Fieser, Louis F. , and Fieser, Mary. Style Guide for Chem-
       ists. New York: Reinhold, 1960. Index. 116p.
This style guide began as a 1945 compilation of guidelines for contributors to Organic Reactions. The advice is taken from such authorities as Fowler, Nicholson, Evans, and Strunk and White. The
first 12 chapters cover conciseness, coherence, verbs, singular and
plural forms, possessives, emphasis, word choice, punctuation,
style, spelling and abbreviations, proofreading, and pronunciation.
Chapter 13, "Speaking, " tells how speaking is different from writing.

268 Fischman, Burton L. Business Report Writing. Providence,
       R.I.: PAR, 1975. Index. 188p.
This text-workbook with tear-out pages has 16 chapters on such

topics as the purpose of the report, clarity, research, short reports, formal reports, memoranda, and problems in report writing.   Forty-four writing exercises are interspersed throughout the chapters.

269   Fishbein, Morris.   Medical Writing: The Technic and the Art.
          4th ed.   Springfield, Ill.: Thomas, 1972.   Index.   203p.
This book is intended for physicians who wish to submit articles to medical journals.   The author, a doctor and long-time medical editor, offers practical advice on such topics as style, words and phrases, spelling, pharmaceutic products and prescriptions, illustrations, revision and proofreading.   Useful tips are provided on writing and preparing manuscripts.   The chapter on indexing, for example, suggests using colored pencils and sets of pencil strokes to indicate indentation of subentries and sub-subentries.   All illustrations and examples are from medical sources.   The author emphasizes correct usage of medical terms and phrases.

270   Flesch, Rudolf.   How to Write Plain English: A Book for Law-
          yers and Consumers.   New York: Harper and Row, 1979.
          Index.   126p.
This book applies the author's well-known readability formula to consumer regulations, primarily those of the Federal Trade Commission.   The 11 chapters cover such topics as "Learning Plain English, " "How to Find Examples, " and "The Cross-Reference Habit." Chapter 2, "Let's Start with the Formula, " gives a detailed explanation of how to compute the formula and includes a chart which will shorten the process.   Chapter 5, "How to Write Plain Math, " discusses methods for writing clear descriptions of mathematical steps. Each chapter is made up largely of examples from government agencies and the revisions that meet the Plain English standards.

271   Flesch, Rudolf.   Rudolf Flesch on Business Communications:
          How to Say What You Mean in Plain English.   New York:
          Barnes and Noble, 1974.   Index.   163p.
Published in hardcover by Harper and Row as Say What You Mean (1972), this book has 12 chapters under such titles as "Get the Facts, " "Use Short Sentences, " "Use Short Words, " and "Explain!" that offer advice on developing clear and concise writing.   Many of the examples come from business.   Chapter 8, "React!" offers ten principles for answering business letters.   Chapter 12, "Keep It Up, " summarizes the main points of the book and suggests that readers compare their own letters with examples in the book.

272   Floyd, Elizabeth R.   Preparing the Annual Report.   AMA Re-
          search Study 46.   New York: American Management As-
          sociation, 1960.   112p.
This study is in three parts.   Part I, "Steps in Preparing the Annual Report, " contains advice on preparation, writing, and distributing the annual report.   Part II, "Content of Annual Reports, " covers the organization and selection of information to be included in the annual report.   Part III, "Exhibits, " illustrates sections from various actual annual reports.

273   Folz, Roy G.   Management by Communication.   Philadelphia:
         Chilton,   1973.

274   Foster, John, Jr.   Science Writer's Guide.   New York: Co-
         lumbia University Press, 1963.   253p.
The first part of this guide discusses "Principles and Practices" with
advice about careers in science writing, writing tips, and collabora-
tion techniques.   The second part (206 pages) is "Definitions of Contemporary Scientific Terms, " covering abampere to zygote.   In the
selection of terms, "the emphasis is on the sciences of great importance for today and tomorrow--modern physics, biology, genetics,
psychiatry, astronomy, and oceanography. "   The third part is a six-
page "Guide to Medical Terms, " showing prefixes, suffixes, and
root words for medicine.   An appendix covers "Masses and Decay
Properties of the Elementary Particles. "   Included is a list of important scientific journals.

275.   Fox, Rodney.   Agricultural and Technical Journalism.   West-
         port, Conn.:   Greenwood, 1976.

276   Frailey, Lester Eugene.   Effective Credit and Collection Let-
         ters.   New York: Prentice-Hall, 1941.   Index.   414p.
This is "a work manual rather than a textbook" for "credit men
everywhere. "   The first three of 14 chapters cover the nature of
collections and credit problems: the function of the "credit man, "
the essentials of the credit letter, and the nature of the people who
owe creditors.   The next several chapters cover the collection cycle.
One chapter covers appeals: self-interest, legal action, etc.   Final
chapters cover unique strategies and situations: showmanship, use
of special days and events, use of collection letters to increase
sales.   The last chapter illustrates how certain companies use the
collection cycle.

277   Frailey, Lester Eugene.   Handbook of Business Letters.   rev.
         ed.   Englewood Cliffs, N. J.:   Prentice-Hall, 1965.   In-
         dex.   918p.
This is a practical handbook directed to people on the job.   The book
brings together advice on building goodwill, constructing clear sentences, and putting together effective paragraphs.   The first six
sections (almost half the book) are devoted to communication principles, business style, format, mechanics, and the expression of
company personality in letters.   The last eight sections cover the
variety of business letters.   There are numerous sample letters.

278   Frailey, Lester Eugene.   Sales Manager's Letter Book.   New
         York: Prentice-Hall, 1951.   Index.   496p.
This book offers hundreds of model letters with advice on writing
them.   There are six major sections with four to eight subdivisions
in each section.   The letters are real samples.   Section I provides
general strategy for writing sales letters, including mistakes to
avoid.   Section II, "Supplementary Sales Letters, " covers letters
that precede or follow sales letters: after a sales call, apprecia-

tion for orders, etc.   Section III illustrates "Letters to Regain Lost Buyers" and "Letters to Win Goodwill" (holidays, condolence, etc.). Section V, "Human Relations in Letters," covers letters designed to motivate employees and build goodwill within a company.   Section VI, "Sales Manager Letters," covers management problems and letters/ memos that help solve them.

279   Frailey, Lester Eugene.   Smooth Sailing Letters.   New York: Prentice-Hall, 1938.   Index.   171p.

This book is essentially a guide to the principles of good business correspondence: word choice, effective format and stationery, good organization, good beginnings, coherent and complete content, strong closings, and courtesy.   The appendix, "Proof of the Pudding," contains fifty examples of good letters.   The author states that the book provides "common sense--served on a plain platter without sauce."

280   Frailey, Lester Eugene, and Schnell, Edith L.   Practical Business Writing.   New York: Prentice-Hall, 1952.   Index. 697p.

The first two of 16 chapters in this book discuss the purpose and process for writing letters.   Chapters 3-4 cover personality and tone.   Chapter 5 discusses the appearance of letters, stationery, and envelopes.   Chapter 6 discusses dictation and the "Business Correspondence Team" ("The Secretary Co-operates with Her Boss" is one section).   Chapters 7-15 cover various types of letters (392 pages).   Chapter 16 discusses business reports (24 pages) and is followed by an index of contributors.   The book contains exercises and discussion questions.

281   Franco, Leonard N., and Zall, Paul M.   Practical Writing in Business and Industry.   North Scituate, Mass.:   Duxbury, 1978.   264p.

Part I of this book, "What You Will Be Writing," covers reports, memos, and letters.   Part II, "The Troubleshooter's Guide," is a reference on the writing process, style, and mechanics.   There are writing-assignment exercises, and Part II uses military numbering. There are three appendixes: "Business Letter Formats," "Sample Technical Proposal," and "Technical Report."   To give the reader "a sense of what it is like to write in a business-industrial environment, with one eye on budgets and the other on schedules, your mind on the problem at hand, and your emotions tied up in the act of writing," the authors have created the fictional company "Bellco."   The examples used throughout are based on problems within Bellco.

282   Freeman, Joanna M.   Basic Technical and Business Writing. Ames: Iowa State University Press, 1979.

283   Freeman, Thomas Walter.   The Writing of Geography.   Manchester, England: Manchester University Press, 1971. Index.   91p.

This book is a guide for writing a thesis in geography.   There are five chapters covering the process of research and writing.   The last chapter deals with clear style.

284   Fruehling, Rosemary, and Bouchard, Sharon.   The Art of Writ-
         ing Effective Letters.   New York: McGraw-Hill, 1972.
         257p.
This textbook covers such topics as organization and coherence, force-
fulness and clarity, the "you attitude," written requests, sales, collection, credit letters, and job applications.   There is a reference
section at the end.

285   Fruehling, Rosemary T., and Bouchard, Sharon.   Business
         Correspondence/30.   2nd ed.   New York: McGraw-Hill,
         1976.   170p.
This workbook is designed to guide high school and two-year college
students in theory of business writing, planning, writing effective
sentences and paragraphs, and concern for the reader.   Included are
specific types of letters, along with résumés and short reports.
Each of the 15 chapters begins with a discussion of the topic, followed
by sample letters for analysis and exercises to reinforce the writing
principles.   Most of the chapters dealing with specific letter types
include checklists so that students can evaluate their own writing.
A final reference section summarizes basic points of punctuation,
abbreviation, and styles of numbering and capitalization.   It also
includes four sample letters showing different typing styles.

286   Gabard, E. Caroline, and Kenny, John P.   Police Writing.
         Springfield, Ill.:   Thomas, 1957.   Index.   93p.
This handbook covers in ten chapters the three basic types of police
literature: scholarly papers (term papers, theses, dissertations),
the journalistic articles, and reports.   Chapters 1-4 cover effective
and clear style, manageable scope, title, and organization.   Chapter
5 lists the functions of the Master's thesis and discusses the role
of the committee, university librarian, and typist.   Chapter 6 focuses
on the police article for publication.   Chapters 7-9 cover the research process, footnotes, etc.   Chapter 10 covers visuals.

287   Gallagher, Oscar C., and Moulton, Leonard B.   Practical
         Business English.   Boston: Houghton Mifflin, 1918.   In-
         dex.   226p.
This high school text covers business English both as a style and as
a kind of writing aimed at making "some sort of appeal that will in-
duce one person to act as another person wishes...."   Its 141 sections with exercises are divided into four parts.   Part I defines
business English and illustrates its use with the four traditional
modes of discourse.   Part II covers writing style, oral presentation, punctuation and letter mechanics, and some forms of business
writing, such as advertising and applications.   Part III shows standard principles of sentence structure and grammar.   Part IV covers
types of letters, including a "jocular collection letter."

288   Gallagher, William J.   Report Writing for Management.   Read-
         ing, Mass.:   Addison-Wesley, 1969.   Index.   216p.
The 12 chapters of this book include "Developing the Input," "Organizing the Report," "Reviewing for Conciseness," "Publishing the Report."   Chapter 10 covers precise grammar points, and Chapter 12
covers preparation of copy, use of color, graphics, etc.

289 Gallagher, William J. So You Have to Write a Report. Cam-
        bridge, Mass.: Arthur D. Little, 1963. Bib. 39p.
This book, written in an informal style and illustrated with cartoons,
is designed for accountants who must write reports for clients. Chapter 1, "Facing the Facts," covers attitudes about writing and writing
for the reader. Chapter 2, "Organizing the Material," deals with
outlining and methods of development. Chapter 3, "Writing the First
Draft," explains paragraph development, transition, and using tables
and graphs. Chapter 4, "Revising the Draft," discusses accuracy
and common grammar problems. Chapter 5, "Mechanics," covers
punctuation, use of numbers, and citing references.

290 Gammage, Allen Z. Basic Police Report Writing. 2nd ed.
        Springfield, Ill.: Thomas, 1974. Index. Bib. 327p.
This book of 23 chapters aims at providing police officers with skills
in the fundamentals of English and in report writing. Part I, "Introduction," is devoted to acquainting the officer with the purposes,
values, and principles involved in report writing. Part II, "Mechanics of Report Writing," is a manual to improve diction, spelling,
capitalization, punctuation, and other basics. Three chapters cover
organization and editing of sentences and paragraphs. Part III, "Reporting Police Operations," deals with preparation of forms used in
police reporting. One section covers "Field Note-Taking." The
reader is taken through the entire process from the recording of an
incident to the specialized forms used in reporting investigations,
arrests, identification, etc. Three appendixes are "Highway Patrol
Forms," "Uniform Classification of Crimes," and "Spelling the
Nouns." Illustrations are of actual department forms and reports.

291 Gammage, Allen Z. Study Guide for Basic Police Report Writ-
        ing. Springfield, Ill.: Thomas, 1975. 288p.
This book is a pedagogical supplement to Basic Police Report Writing. Each chapter here corresponds to a chapter in the other book.
All chapters contain a summary, outline questions, terms and concepts, chapter quiz, assignments, and answers to quiz and exercises.

292 Gardner, Edward Hall, and Aurner, Robert Ray. Effective
        Business Letters. rev. ed. New York: Ronald, 1928.
        Index. 385p.
This book, originally published in 1915, was based on a course,
"Business Letter Writing," at the University of Wisconsin. It focuses
on communication and the general aims of business. There are 23
chapters. Chapters 1-2 discuss persuasion and image building in
business letters. Chapters 3-21 cover types of letters (credit, sales,
inquiry, etc.) and the persuasive strategy for each. Chapter 22
covers developing mailing lists, and Chapter 23 reviews business
reports. Chapters end in case problems as well as correction sentences.

293 Gardner, G. H. Constructive Dictation, "Plan Your Letter."
        Chicago: Gregg, 1919.

294 Garn, Stanley M. Writing the Biomedical Research Paper.

Springfield, Ill.: Thomas, 1970. 65p.
This book is designed to help "scientist-authors, not necessarily only
beginners" prepare research reports for biomedical journals. Topics
include report title, author identification, abstract, introduction,
methods and materials, findings, discussion, bibliography, and references. The book also advises on use of tables, charts, illustrations, legends, and acknowledgments. The last four chapters discuss such production and publication aspects as mailing manuscripts,
the review process, reading and returning proofs, and reprints.

295 Gensler, Walter J., and Gensler, Kinereth D.   Writing Guide
        for Chemists.   New York: McGraw-Hill, 1961.   Index.
        149p.
Written for students and professionals, this book has 15 chapters in
two parts.   Part I covers elements of technical writing for chemists:
style, laboratory notebook, written outline, elements of the report,
effective presentation of details, and revision.   Part II covers such
structural and mechanical problems as inappropriate expressions,
ambiguous grammatical relationships, punctuation and italics, spelling, and capitalization.   This part also covers such special topics
as chemical nomenclature, forms of physical data, structural formulas, and documentation of chemical journals.

296 Gibbs, Helen M.   The Research or Technical Report: A Manual
        of Style and Procedure.   San Francisco: Robert R. Gib-
        son, 1950.   Index.   Bib.   131p.
This guide of ten chapters is set in typewriter font.   Topics covered
include the parts of the report, punctuation, capitalization, the hyphen, symbols and numbers, and documentation.   Chapter 8 explains
how to cite foreign and American currency.

297 Gieselman, Robert D., ed.   Readings in Business Communica-
        tion.   2nd ed.   Champaign, Ill.: Stipes, 1978.   254p.
This collection presents 25 articles.   Selections include "Written and
Spoken English for the Accountant," by Robert H. Roy and James H.
MacNeill; "The Fog Index After Twenty Years," by Robert Gunning;
and "Reducing Sexually Biased Language in Business Communication,"
by Daphne A. Jameson.

298 Gilbert, Marilyn B.   Letters That Mean Business.   New York:
        Wiley, 1973.   Index.   256p.
This book of nine chapters is a programmed self-instruction guide.
Chapter 1 covers effective paragraphing for letters.   Chapters 2-4
cover the letter patterns the author considers basic: asking, telling,
and building goodwill.   Chapters 5-6 discuss editing various problems: affectation, wordiness, passive voice, and expletives.   Chapter 7 covers copy editing for common errors, numbers, dividing
words, and spacing.   Chapter 8 covers letter format, and Chapter
9 covers the résumé.   Following a "Closing Remarks" section is a
75-page reference handbook with self-tests, advice on writing telegrams, sample letters and résumés, and letter outlines.   Also included are forms of address, abbreviations, spacing information,
and a glossary of business terms.

299 Gilman, William.   The Language of Science: A Guide to Ef-
          fective Writing. New York: Harcourt, Brace and World,
          1961. Index. 248p.
This book is an informal discussion of the need for clear writing.
The 15 chapters cover such topics as "The Answers to Jargon,"
"Hyphen Horrors," "Sentences That Move Faster," and "The Second
Look." The numerous illustrations are from published and unpub-
lished sources. The author stresses that clarity in technical writing
does not make it "pop" but does make it readable.

300 Glendinning, Eric H.   English in Mechanical Engineering.   Lon-
          don: Oxford University Press, 1974. 131p.
This textbook for non-native speakers of English presumes that the
students have "a considerable dormant competence in English" and
"a basic knowledge of their specialist subject." The eight units cover such topics as "The Four-Stroke Petrol Engine," "Levers," and
"Friction." Each unit is divided into five sections. Section I,
"Reading and Comprehension," is a passage on the topic with true-
false questions. Following are solutions to the questions, showing
the student the reasoning process needed for each question. Three
exercises follow, covering rephrasing, pronoun reference, and transition. Section II, "Use of Language," contains exercises based on
charts and diagrams and sentence practice. Section III, "Information
Transfer," contains exercises based on those in Section II. Section
IV, "Guided Writing," provides practice in sentence and paragraph
building and writing tasks. Section V, "Free Reading," is a passage
that is longer and more difficult than the passage in Section I.

301 Glidden, H. K.   Reports, Technical Writing, and Specifications.
          New York: McGraw-Hill, 1964. Index. 312p.
This textbook has 15 chapters divided into three parts: Part I, "Reports and Industry," covers the nature and content of reports; Part
II, "Writing the Report," deals with the writing process, research,
style, format, and editing; Part III, "Special Types of Writing,"
covers articles, manuals, oral reporting, and specifications. Appendix A presents proofreader's marks, and Appendix B covers giving
credit and receiving copyright permission. Each chapter ends in
writing/researching exercises.

302 Gloag, John.   How to Write Technical Books--With Some Perti-
          nent Remarks on Planning Technical Papers and Forms.
          London: Allen and Unwin, 1950. Index. Bib. 159p.
This book, designed for British technical authors, advocates a clear
and readable style for books on technical subjects. It also offers
advice and information about producing technical books from the in-
ception of the idea and planning to dealing with a publisher and book
production. The author also devotes a chapter to "Technical Docu-
ments--papers, reports and memoranda." The final chapter of the
ten offers suggestions on improving standard forms in business and
government, which (according to the author) have just "happened"
rather than been designed. The appendixes are two bibliographies:
one on book planning, the other on general style and usage reference.
The book is illustrated with photographs, and drawings, and color
reproductions of forms.

303 Glover, John George. Business Operational Research and Re-
     ports. New York: American Book Company, 1949. In-
     dex. 299p.
This book covers conducting research in business, science and technology, and writing about the results. The eight chapters concen-
trate on research, organization, and physical format. Chapter 8
covers writing the report. There are five appendixes covering almost as many pages as the main text. Appendix A offers "Appropriate Subjects for Research and Study," an extensive subdivided
list. Appendix B, "The Preparation of Reports," is a reprint of a
booklet published by Hercules Powder Company in 1945. Appendix
C, "Hints to Authors: Preparation of Mss. for the National Industrial Conference Board," reprints the board's guidelines for authors.
Appendix D, "A Method of Outlining the Academic Research Thesis,"
is a sample outline without supplementary comment. Appendix E
is a "Sample of Title Page of Academic Thesis."

304 Godfrey, J. W., and Parr, G. The Technical Writer. Lon-
     don: Chapman and Hall, 1959. Index. Bib. 340p.
This book is designed to give the British technical writer "insight
into the technique of presentation and production" of technical literature. Chapters 1-4 cover forms of technical writing, readership,
research, organization, style, usage, grammar, and illustrations.
Chapters 5-8 cover the printing process, typesetting, editorial procedure in layout and proofs, typeface, and binding methods. Chapter
9 makes suggestions for setting up a technical-publications department. Chapter 10 covers writing the technical book. The four appendixes include three sections on printer terms and publishing information and one section on associations for technical writers.

305 Goeller, Carl. Writing to Communicate. New York: New
     American Library, 1974. Index. 142p.
This book is designed to teach business people to communicate "exactly" what they have in mind. The 11 chapters cover in an informal style clarity, accuracy, and brevity in letters, memos, reports,
speeches, and job applications. One chapter provides advice for
writing articles, papers, and books. The last chapter gives a 30-
day program for improving writing, editing, and logical thinking.
A special feature of the book is its use of whole letters that are
marked with editing changes.

306 Goodman, David S. President's Letter Book. Englewood
     Cliffs, N. J. : Prentice-Hall, 1970. 311p.
A binder with colored tabs marking sections, this book provides
model letters for the executive secretary or administrative assistant
to whom the executive gives most writing tasks. The introduction
stresses general letter-writing principles of diplomacy, preciseness,
getting results, and brevity. The reader is also cautioned about
verbosity and gobbledygook. Chapters include "Letters to Internal
Management," "Letters to Employees," "Letters on Financial Matters," "Letters to Customers," "Letters to Media," and "Letters
to Suppliers." Each section begins with a brief statement about the
type of letter. Sample letters are shown on the right-hand pages
with the author's commentary on the left-hand pages.

307 Gould, Jay R. , ed. Directions in Technical Writing and Com-
          munication. Farmingdale, N. Y. : Baywood, 1978. 152p.
This collection of 14 articles is divided into five parts: "What Is
Technical Communication?" "Basic Forms of Technical Communication, " "Technical Communication in Practice, " "Evaluating Technical
Communication, " and "Viewpoints in Technical Communication. "
Articles include "External Examiners for Technical Writing Courses, "
by Thomas M. Sawyer; "The Persuasive Proposal, " by Lois DeBakey; and "The Trouble with Technical Writing Is Freshman English, "
by W. Earl Britton. All the articles were originally published in
the Journal of Technical Writing and Communication.

308 Gould, Jay R. Practical Technical Writing. Washington, D. C. :
          American Chemical Society, 1973.

309 Gould, Jay R. , and Losano, Wayne A. Opportunities in Tech-
          nical Writing Today. Louisville, Ky. : Vocational Guid-
          ance Manuals, 1975. Index. Bib. 137p.
This book for vocational counselors has seven chapters covering the
opportunities and requirements for success in technical writing. Included are such topics as the importance of the professional today,
educational requirements, schools with degree programs in technical
communication, kinds of employers who need technical writers, employment outlook, and the way to get started.

310 Gould, Jay R. , and Olmstead, S. P. Exposition: Technical
          and Popular. New York: Longmans, Green, 1947. 126p.
This book is designed as a supplement in "courses where little time
is ordinarily devoted to expository writing. " The book has seven
chapters in three sections. Section I, "Forms of Expository Writing, " covers definition, instructions, descriptions of mechanisms,
processes, and general exposition of argumentation. Section II covers only the research article. Section III covers adaptation of technical material for the general reader.

311 Graham, J. , and Oliver, G. A. S. Foreign Traders' Corres-
          pondence Handbook; For the Use of British Firms Trading
          with France, Germany and Spain, Their Colonies, and
          with Countries Using Their Languages. New York: Mac-
          millan, 1905.

312 Grantham, Donald, et al. Technical Communication. Los An-
          geles: GSE, 1975.

313 Graves, Harold F. , and Hoffman, Lyne S. S. Report Writing.
          4th ed. Englewood Cliffs, N. J. : Prentice-Hall, 1965.
          Index. 286p.
This text, first published in 1929, stresses the need for the writer
to be "audience-minded. " The first four chapters cover the importance and the demand for reports, letters (inquiry and instruction)
and the memoranda, general style requirements, and the collection
of data, including observation and experiment. Chapters 5-6 cover
planning the report, outlining, and parts of the report and their

rhetorical purpose.   Chapter 7 covers format and documentation.
There is a section that provides specimen reports and a 54-page
handbook of grammar and usage.   Appendix A discusses the letter
of application, and Appendix B is a "Bibliography of Abstracts and
Indexes in technical fields. "

314   Gray, Dwight E.   So You Have to Write a Technical Report.
        Washington, D. C. :   Information Resources, 1970.

315   Gresham, Stephen L. ; Rivers, William; and Waltman, John L.
        Letter Writing for Public Officials.   Auburn, Ala. :   Au-
        burn University Office of Public Service and Research,
        1977.   Index.   152p.
This book is devoted to guiding public officials in writing readable
prose.   Chapters 1-3 cover effective style, placement of information,
and reader orientation.   Chapter 4 is a short handbook, "Grammar,
Mechanics and Usage. "   Chapter 5 covers relevant types of correspondence.

316   Griffin, Margaret P.   Practical Approach to Communicating in
        Writing and Speaking.   New York: Glenco, n. d.

317   Grinsell, Leslie; Rahtz, Philip; and Williams, David Price.
        The Preparation of Archaeological Reports.   2nd ed.
        London: Baker, 1974.   Index.   Bib.   105p.
This book is for the professional writing reports on archaeological
findings.   The three authors have written the seven chapters inde-
pendently; each chapter has a byline.   Chapter 1, "The Form of
Publication" is a discussion of the difficulties and costs of publishing
and under what circumstances the archaeologist should publish.   Chapters 3-4 deal with excavational reports; Chapter 5 covers nonexcava-
tional reports.   Appendix A is a chart of archaeological sites and
the reference books where artistic reconstructions can be found.   Appendix B is an overview of publications recommended for study.
There are ample illustrations of the kinds of drawings for inclusion
in reports.

318   Grinter, Linton E.   Writing a Technical Report.   New York:
        Macmillan, 1945.

319   Guidelines for Research in Business Communication.   Urbana,
        Ill. :   American Business Communication Association,
        1977.   96p.
This publication, prepared by the Research Committee of ABCA, is
a collection of ten articles on research in business communication.
Part I, "Planning and Conducting Research, " includes articles on
laboratory, field, and historical research.   Part II, "Analyzing Research Data, " includes an article on computer applications and how
data are measured.   Part III, "Reporting Research Findings, " includes articles on arrangement and content of research reports.   A
short bibliography appears at the end of each article.   The purpose
of this collection is to offer assistance in promoting and conducting
quality research among the ABCA members.

320 Gunning, Robert. New Guide to More Effective Writing in Busi-
            ness and Industry.   Boston: Industrial Educational Insti-
            tute, 1963. 278p.
This book is divided into nine sections.   Section I is a discussion of
the high cost of poor writing and an explanation of the scope of the
book.   The author states, "what you communicate and to whom you
communicate are largely matters of business administration.   I in-
tend to talk only about how to communicate." Section II covers the
general principles of writing and includes a division on the Fog Index.   Section III, the largest, covers organization. Sections IV-V
deal with effective style.   Sections VI-VII cover the writing and editing process.   Section VIII provides short notes on specific kinds of
writing, such as bulletin-board announcements, letters, and proposals.   Section IX provides examples of actual writing from industry
with a Fog count for each.   The appendix provides an excerpt, "Technical Writing," taken from The Technique of Clear Writing, by the
same author.

321 Gunning, Robert. The Technique of Clear Writing. rev. ed.
            New York: McGraw-Hill, 1968. Index. 329p.
This book, originally published in 1952, is divided into three parts.
Part I, "What Your Reader Wants," discusses in three sections the
danger of clouded or "foggy" writing, analyzing the reader, and various readability formulas, including the author's own Fog Index. Part
II, "Ten Principles of Clear Writing," covers style and reader analysis.   Part III, "Causes and Cures," covers newspaper, business,
legal, and technical writing.   The book concludes with three appendixes: a readability analysis of the Gettysburg Address, the Edgar
Dale List of 3,000 Familiar Words, and a list of long words with
short-word alternatives.

322 Gunning, Robert. What an Executive Should Know About the
            Art of Clear Writing. Chicago: Dartnell, 1968.

323 Guthrie, Ledru O. Factual Communication: A Handbook of
            American English. New York: Macmillan, 1948. Index.
            Bib. 448p.
This text is designed for "directly practical uses of English in articles, talks, business letters, and reports." Aimed at students in
science, engineering, and business, the book has nine chapters divided into three parts.   Part I includes chapters on the writer's
relationship to the message and reader, style, outlining, illustrations, making a factual talk, methods of development, types of reports, and business letters.   Part II covers clarity, correctness,
and accuracy; the writing and editing process; and research methods.
Part III is a handbook for sentence construction and variety.   The
last chapter is on grammar.

324 Hagar, H. A. Applied Business English. Chicago: Gregg,
            1909.

325 Hagar, Hubert A., et al. Business English and Letter Writing.
            New York: McGraw-Hill, 1953.

326 Hall, H. L. C. Commercial Correspondence. Detroit: Book-
     keeper, 1906.

327 Hall, Mary. Developing Skills in Proposal Writing. 2nd ed.
     Portland, Ore.: Continuing Education Publications, 1977.
     339p.
This book, set in typewriter font, is based on the assumption that
"there is no special mystique about proposal writing." The author
regards proposal development as a process. Part I, "The Prepro-
posal Phase," contains four chapters covering the process of getting
ideas, assessing the competition for funds, developing the idea and
selecting the funding source. Part II, "The Proposal Phase," has
ten chapters focusing on the writing process. Chapter titles include
"Writing the Proposal," "Writing the Statement of Need," "Writing
Objectives," and "Writing Procedures." Also covered is information
on dissemination, personnel, and budget. Included are samples of
forms, model letters, budgets, worksheets, and proposal criteria
from several funding sources. The appendix is a 13-page "Proposal
Development Checklist."

328 Hall, Samuel Roland. Business Writing. New York: McGraw-
     Hill, 1924. Index. 222p.
This book, based on the author's experience in business departments
of newspapers, has six chapter-length sections: "Business-Magazine
Articles," "Newspaper Items," "Copy for House Organs," "Writing
a Report," "Copy for Advertisements," and "Improving Your English." The section on newspaper items really covers news releases
on business items. There are illustrations of contemporary articles
and ads.

329 Hall, Samuel Roland. Handbook of Business Correspondence.
     New York: McGraw-Hill, 1923.

330 Hallock, Virginia L. Business Communication: Effective Cor-
     respondence Through the Tri-Ask Technique. Providence,
     R.I.: PAR, 1974.

331 Hammond, H. W. Style-Book of Business English Designed for
     Use in Business Courses, Regents' and Teachers' Exami-
     nations. 5th rev. ed. New York: Pitman, 1913. In-
     dex. 232p.
Geared for use in schools where no teacher of Business English is
employed, the 32 chapters in this text are divided into three parts.
The author points out that new to this edition is a section on private
secretaries' duties and a section on filing letters. Part I contains
chapters on specific types of letters, tone, social correspondence,
forms of address, the paragraph, and punctuation. Part II is intended for use in shorthand and typing classes. It includes proofreading marks, foreign words in business, frequently misspelled
words, postal information, telegrams, and cablegrams. Part III
explains card indexing and letter filing. One chapter covers abbreviations and the exchange value of foreign coins. Another chapter
advises teachers on the presentation of business-English principles.
There are short answer exercises.

332 Hammond, Kenneth R., and Allen, Jeremiah M., Jr. Writing
Clinical Reports. New York: Prentice-Hall, 1953. Bib.
235p.
This book is written primarily for professional psychologists and for
graduate students. The eight chapters cover such topics as the primary and secondary purposes of psychological reports (for clinical
use and scientific reference), primary and secondary readers, organization and style, and adapting technical vocabulary. Chapter 5
covers readability and tone, including a discussion of avoiding affectation and maintaining objectivity. Chapter 8 covers good revising
and editing practices. The four appendixes give samples of printed
forms for patient evaluation, sample reports, a vocabulary test for
the reader, and a brief, annotated list of reference books in language, usage, and psychology.

333 Handy, Ralph S. Business Correspondence in Practice. 3rd
ed. New York: Pitman, 1962. 252p.
This workbook presents 25 chapters for a complete course in business
letters. The illustrative letters are taken from files of actual business firms. The pages are detachable and punched so they can be
placed in a binder. Topics covered include format, folding and
stuffing the envelope, language, tact, cheerful tone, friendliness,
types of letters, and punctuation. Most exercises are fill-in, short
answer, or true-false. Writing assignments ask for a specific letter, e.g., write to a clergyman who is in arrears on his motorcycle
payments.

334 Harbarger, Sada A., et al. English for Students in Applied
Sciences. New York: McGraw-Hill, 1938. Index. Bib.
260p.
This textbook is designed for beginning students in schools of engineering and applied sciences. The 14 chapters cover rhetorical
principles but emphasize their application to technical writing. Professional standards in technical communications are illustrated by
letters, reports, articles, and excerpts from books by competent
writers in applied science. Part I is devoted to writing principles,
organization and outlining of reports, "figures and tables," and the
production of "The Final Copy." Part II deals with applications of
these principles to different kinds of reports, research (formal) and
informal. The three appendixes deal with reading lists for students
and selected references for teachers and students. Writing assignments are included.

335 Harris, J. R., et al. Technology for the Formulation and Expression of Specifications. 3 vols. Urbana: University
of Illinois Press, 1975. Index. Bib. 235p.
This 12-chapter report was sponsored by the National Bureau of
Standards and carried out by the Departments of Civil Engineering
and Architecture of the University of Illinois. The "report describes
a systematic approach to the formulation and expression of specifications ... designed to aid in producing complete, clear, and correct
documents." "The approach primarily is concerned with the format
of specifications" and provides models for construction specifications.

336 Harris, John S. , and Blake, Reed H. Technical Writing for
     Social Scientists. Chicago: Nelson-Hall, 1976. Index.
     122p.
This book is designed to deal with the special problems of writing in
the social sciences, e. g. , multiple words for the same concept. The
book is divided into six major sections: "First Considerations"
(audience, purpose, scope, plan), "Writing the Sentence and Paragraph, " "Defining the Problem" (hypothesis development and literature review), "Gathering Information, " "Organizing the Data, " "Writing the Paper: Style, Format, and Language. "

337 Harrison, James, ed. Scientists as Writers. Cambridge,
     Mass. : MIT Press, 1965. Index. Bib. 206p.
This "selection of prose passages by scientists about science" is
intended for use in "courses where it is hoped the students will
sharpen their wits, improve their command of English, and at the
same time broaden the over-specialized field of study covered by
their other subjects. " Most of the selections in the 11 chapters are
excerpts from larger works on the nature of science, discoveries,
and the relation of art to science. One section provides chapter-by-
chapter suggestions for in-class discussion and writing. Following
this section are "Suggestions for Further Reading" and "Biographical
Notes. "

338 Hart, Andrew W. , and Reinking, James A. Writing for the
     Career-Education Student. New York: St. Martin's,
     1977. Index. 335p.
Designed especially for students in vocational and technical programs,
the book has 12 chapters, moving from a general overview of writing
to an increasingly technical emphasis. Chapter 1 discusses writing
the paragraph, and Chapter 2 takes the student through the steps of
preparing a theme. The next five chapters cover comparison, classification, explanation of a process, definition, and description of an
object. Chapter 8 covers letters and memos. Chapters 9-11 deal
with proposals, progress reports, and investigation (or text) reports.
The text contains writing exercises and some discussion questions.

339 Harty, Kevin J. Strategies for Business and Technical Writing.
     New York: Harcourt Brace Jovanovich, 1980. Bib.
     285p.
This anthology contains 25 articles that are divided into five sections.
The sixth section is an annotated bibliography of books, articles, and
bibliographies. Section I contains seven articles on general communication principles. Section II contains three articles on the problem
of jargon, including "A Rationale for the Use of Common Business-
Letter Expressions, " by J. Harold Janis. Section III contains two
articles about job applications. Section IV provides six articles on
letters and memos, including "The Legal Aspects of Your Business
Correspondence, " by Herta A. Murphy and Charles E. Peck. Section V has seven articles on reports, including "Audience Analysis:
The Problem and a Solution, " by J. C. Mathes and Dwight W. Stevenson, and "The Writing of Abstracts, " by Christian K. Arnold.

340   Harvill, Lawrence R., and Kraft, Thomas L. Technical Report
          Standards: How to Prepare and Write Effective Technical
          Reports. Sherman Oaks, Calif.: Banner Books Inter-
          national, 1977. Bib. 55p.
This guide covers the presentation of technical data. The six chapters deal with parts of a report, format, tables, graphs, illustrations, and presentation of errors. Each chapter is divided into topics with military numbering. The emphasis is on visual data. Chapter 6, "Errors," discusses the need for including in reports a discussion of possible errors. The chapter also defines types of errors. Included are illustrations of types of charts and graphs.

341   Harwell, George C. Technical Communication. New York:
          Macmillan, 1960. Index. 332p.
Written primarily for the engineering student, this text has ten chapters. Chapters 1-3 cover the qualities of good writing, organization
of material, and the use of exposition. Chapter 4 covers the business letter. Chapters 5-7 treat reports. Chapters 8-9 cover the
magazine article and public speaking. Chapter 10 deals with the use
of and the preparation of tables and figures. The last part is a
handbook of grammar and usage. The exercises cover writing assignments, cases, and class discussion.

342   Hatch, Richard. Communicating in Business. Chicago: Sci-
          ence Research Associates, 1977. Index. 348p.
The basis for this textbook is to present business communication in
the context of "psychological principles of communication." The 14
chapters cover such topics as good-news messages, revising the
draft, the psychology of persuasion, and writing the proposal. Chapter 6, "The Media of Business Communication," contains an analysis
of the cost, speed response, factors in both talking/listening media
and writing/reading media. There are four sections of writing
cases: "Brief Informational Messages," "Bad-New Messages,"
"Persuasive Messages," and "Organizing and Writing a Report."
There are two continuing cases requiring a variety of responses.
There are two appendixes: one covering typing format for the business letter and memo, and the other presenting a seven-page review
of grammar.

343   Hawkey, M. English Practice for Engineers. London: Long-
          man, 1970.

344   Hawkins, Clifford. Speaking and Writing in Medicine: The Art
          of Communication. Springfield, Ill.: Thomas, 1967.
          Index. 159p.
This book aims to help the physician attain better communication with
patients and peers. The first three of 19 chapters cover speaking.
Chapter 4, "Clear English," discusses the need for directness in
scientific writing and compares scientific with literary style. Chapter 5 examines the writing and publication process for medical articles. Chapter 6 focuses on obscurity, jargon, and readability.
Chapter 7 covers the use of illustrations. The final two chapters
review doctor-patient communication (listening and talking). A

three-part appendix covers "Proofreader's Marks, " "Common Abbreviations and Symbols, " and "American and English Usage in Spelling. "

345 Hay, Robert D.   Written Communications for Business Admin-
        istrators.   New York: Holt, Rinehart, and Winston,
        1965.   Index.   487p.
This textbook is aimed at those already in administrative positions
as well as the student in business communication.   The 23 chapters
are divided into three parts.   Part I, "Business Correspondence, "
covers the psychology of business letters and reader analysis, tone,
planning, clarity, format, completeness and conciseness, grammar,
and punctuation, and the job search.   Chapter 6, "Planning the Letter, " stresses five basic letter-writing situations that the author
says cover all letter types.   Chapter 11, "Special Problems of the
Manager, " discusses dictation, form letters, mailing, and supervising correspondence units.   Part II, "Report Writing, " covers gathering data, organizing, graphics, and writing special reports (staff-
study, and letter).   Part III, "Employer-Employee Communications, "
discusses principles of such communication and writing policy statements and procedures.   Chapters end in problems for discussion or
writing.

346 Hay, Robert D. , and Lesikar, Raymond V.   Business Report
        Writing.   Homewood, Ill. :   Irwin, 1957.   Index.   352p.
This textbook, designed for business students, has 18 chapters divided into five parts.   Part I serves as an introduction to business-
report-writing principles.   Part II focuses on the report-writing
process: analysis, information gathering, constructing the report,
and interpreting information.   Part III covers report-writing style
with a discussion of honesty and objectivity.   Part IV covers grammar and punctuation, and Part V discusses reports as management
tools.   Following the chapters is a section of additional report case
problems for the student.   The chapters themselves also provide
writing assignments and questions for discussion.

347 Hays, Robert William.   Principles of Technical Writing.   Read-
        ing, Mass. :   Addison-Wesley, 1965.   Index.   Bib.   324p.
This textbook is designed for technical and engineering students and
persons on the job.   The 13 chapters cover all the aspects of writing
reports: gathering the data, presentation and analysis, outlining,
visual aids, writing the first draft, revision and editing, and final
presentation/submission of the report.   There are nine appendixes
including such topics as guidelines for punctuation, numbering, abbreviations.   One appendix provides sample outlines.

348 Hazelet, John C.   Police Report Writing.   Police Science
        Series.   Springfield, Ill. :   Thomas, 1960.   Index.   238p.
This book of 12 chapters covers proper report technique, study of
basic police records, processing of the records, and the Uniform
Crime Reporting Program of the FBI.   The chapters include "Essential Information for Reports, " "Complaint and Investigative Records, "
"Arrest and Identification Records, " "Traffic Records. "   Many illustrations show both officers doing reporting and report forms.   The
author was the chief of police of Lawrence, Kansas.

349 Held, Felix E. Life Insurance Correspondence. New York:
        Life Office Management Association, 1936. Index. 209p.
Although the author refers to this book as a "text" in his preface,
there are no pedagogical devices, and the book seems more sharply
aimed at life-insurance professionals. The author states in the preface that "the mediocrity of correspondence is generally attributed to
carelessness and lack of preparation on the part of the individual
letter writers.... [But] this text considers the problem one of management, and places the responsibility for better business letters on
executives, administrators and department heads." Chapter 1 discusses the problem of poor letters, the benefits of good letters, and
the excuses businesses and writers give for poor letters. The next
four chapters cover form, style, correctness, and models of good
letters. The following three chapters cover various specific types
of internal and external correspondence. Chapter 9 covers the supervision of letter writing, and the final chapter covers reports and
memos. Following are two appendixes. Appendix I is a summary of
the results of a test on language usage in life-insurance letters. The
test was intended "to convince letter writers that their letters lacked
clarity." Appendix II is a set of model letters for "students."

350 Hemmerling, E. M. Business English Essentials. 5th ed.
        New York: McGraw-Hill, 1975.

351 Hemphill, P. D. Business Communications with Writing Im-
        provement Exercises. Englewood Cliffs, N.J.: Prentice-
        Hall, 1976. Index. 278p.
This text with tear-out worksheets has 15 chapters covering such
topics as communication theory, goodwill, routine information letters,
form letters, credit and collection letters, oral communication, job
applications, and reports. Each chapter ends in review questions
and a grammar/style lesson with worksheet. Appendix A contains
a line guide for placement of letters typed in "picture frame placement form" (centered). Appendix B gives punctuation rules for the
period, comma, and the semicolon. Appendix C contains spelling
lists. Appendix D is a list of questions frequently asked during job
interviews, based on material from New York Life Insurance Company.

352 Hemphill, Phyllis Davis. Career English: Skill Development
        for Effective Communication. Englewood Cliffs, N.J.:
        Prentice-Hall, 1980.

353 Henderson, Greta L., and Voiles, Price R. Business English
        Essentials. 4th ed. New York: McGraw-Hill, 1970.
        Index. 218p.
This book consists of text and workbook. The text (white pages) is
divided into five parts. The first four cover parts of speech, grammar and usage, punctuation, and letters, memos, and formal and
informal reports. Part V is on "Planning and Giving a Talk."
There are writing assignments throughout the text. The workbook
section (green pages) consists of fill-in worksheets keyed to the text
topics. There are special tests in a separate booklet available to

the instructor. At the beginning of the text is a "Course Progress
Record, " where the student records the scores received on the worksheets and the special tests.

354 Hendricks, King, and Stoddart, L. A. Technical Writing.
          Logan: Utah State Agricultural College, 1948. Index.
          117p.
The 58 chapters in this text are divided into five parts designed both
for the student and the experienced researcher. Part I, "Preliminaries to Writing, " discusses selecting and limiting a topic, planning
research, establishing bibliography, notes, and outlining. Part II,
"Composition of Manuscript, " discusses all the parts of a long, formal report, including directions for typing manuscripts and theses.
Part III, "Preparation of Manuscript for Printing, " defines printers'
terms and offers advice on paper, cover, typefaces, size of type,
length of line, margins, reproductions, and proofreading. Part IV,
"Mechanics of Construction, " covers the rules of punctuation and
spelling, with short sections on usage and abbreviations. Part V,
"Reference Materials, " lists references for the fields of agriculture,
art, education, engineering, forestry, language and letters, science,
and social science.

355 Henn, Thomas Rice. Science in Writing. New York: Mac-
          millan, 1961. 248p.
This collection of 31 scientific writings is designed in part to closing
the gap between the arts and the sciences today. The selections
cover a wide range of history--showing "the memorable or typical
use of prose in a variety of fields of observation, analysis, specula-
tion, and judgement. " The selections in Part I are arranged in
historical order from "Nature and Medicine" (Pliny) and "A Vision
of the New Science: 'Solomon's House'" (Bacon) to "Some Philosoph-
ical Aspects of Modern Physics" (Max Born). Most of the authors
selected are British. Part II discusses primarily the stylistic qualities of scientific writing; however, it also covers the writing process
in terms of scientific investigation.

356 Heron, Alexander R. Sharing Information with Employees.
          Stanford, Calif. : Stanford University Press, 1942. 204p.
This work is designed to assess the most effective ways to share
information with employees and select the content for company publications and policies. The 24 chapters begin by examining the employees' attitudes toward the company and management. The book
then analyzes the kinds of information employees want and should
receive. Six chapters deal with forms of writing, content, and
strategy. Chapters deal with direct mail to employees, the company
handbook, the house organ, the annual report, the pay insert, and
the bulletin board. The study concludes with an analysis of public
and employee meetings, the grapevine, unions, and the role of line
supervisors.

357 Hess, Karen M. , and Wrobleski, Henry M. For the Record:
          Report Writing in Law Enforcement. New York: Wiley,
          1978. Index. 273p.

This textbook of 11 chapters offers guidelines for report writing at
all levels of police science and gives a combination of theory, grammatical review, and practice in producing professional police reports.
The first three chapters deal with function, characteristics of reports, and the steps involved in gathering the facts through writing
and evaluating/editing of reports.   Chapters 4-10 cover grammar,
punctuation, usage, spelling, and sentence structure.   The last chapter is devoted to review and report writing practice.   The final chapter gives cases of police investigation and asks students to write reports based on these cases.   Since every chapter has a self-test with
answers provided, the book can be used for self-study.

358 Hewitt, Richard M.   The Physician-Writer's Book: Tricks of
        the Trade of Medical Writing.   Philadelphia: Saunders,
        1957.   Index.   415p.
This book, based on the author's 25 years of experience as the editor
of JAMA, is aimed at practicing physicians and researchers who
want to do medical writing.   The book contains 39 chapters, with
advice for the medical writer, the editor of medical manuscripts,
and for those who are engaged in illustrating the text and supplying
the statistical data for medical articles and books.   Part I deals
with the entire composition (book or article).   Part II covers the
construction of "The Paragraph."   Part III is "The Sentence."   Part
IV deals with words and idiomatic expressions, and Part V with
"Tables and Illustrations."   Part VI covers the preparation of manuscripts for the press, and Part VII covers medical writer's ethics.
The 21 appendixes supplement topics covered in some chapters.

359 Heys, Harry.   Preparation and Production of Technical Hand-
        books.   London: Pitman, 1965.   Index.   Bib.   183p.
This book is designed to help the British technical writer prepare a
manual or handbook.   The author also attempts to elevate the technical
writing profession and define the technical writer's role, especially
in the last chapter (10) titled "Founding and Administering a Technical Handbook Department."   The rest of the book treats in detail
estimating, note taking, drafting, editing, commissioning artwork,
printing, binding, and establishing clerical requirements of a technical "handbook" (writing) department.

360 Hicks, Tyler G.   Successful Technical Writing.   New York:
        McGraw-Hill, 1959.   Index.   294p.
This book is designed to give practical advice to engineers, technicians, and technical writers.   Seven of the 17 chapters, for example,
cover writing technical articles for trade magazines.   Two of the
chapters cover writing industrial catalogs and advertising copy.   The
last three chapters focus on writing technical books.   Each chapter
ends with a summary and a list of questions for self-study about the
chapter contents.   The emphasis of the book is on professional technical writing.

361 Hicks, Tyler G.   Writing for Engineering and Science.   New
        York: McGraw-Hill, 1961.   Index.   Bib.   298p.
Written for "technical writers, be they practicing engineers or scientists, professional technical writers, or writing students in all

branches of science and technology, " 16 chapters of this book cover in depth all the major forms of technical writing--reports, articles, papers, manuals, and specifications.   The book also has chapters on grammar and usage, technical sales and news writing, and manuscript presentations.   The exercises and writing assignments for the chapters are gathered in ten pages at the end of the book.   Some of the chapters are arranged in a handbook-like format.

362   Hillman, Howard.   The Art of Winning Government Grants.
          New York: Vanguard, 1977.   Index.   Bib.   246p.
This book covers the entire process of getting government grants. The three parts are "The Six Grant-Seeking Phases, " "Where the Money Is, " and "Information Sources. "   The appendixes are "History and Trends, " "Grant Genesis, " "The Bureaucratic Monster, " "A-95 Clearinghouse Review Process, " "Student Aid, " "Forms of Address, " "Acronyms, " "Types of Grants, " "Glossary. "   The material is comprehensive.   "Phase Three" in Part I covers writing the grant application.

363   Hillman, Howard, and Arabanel, Karin.   The Art of Winning
          Foundation Grants.   New York: Vanguard, 1975.

364   Himstreet, William C. , and Baty, Wayne M.   Business Com-
          munications: Principles and Methods.   5th ed.   Belmont,
          Calif.: Wadsworth, 1977.   Index.   509p.
This textbook is designed to help students respond to "quasi-realistic" business situations through a "critical, sentence-by-sentence analysis of written communication. "   The 20 chapters of the text are divided into five parts: "Communication Foundation and Writing Principles, " "Communication Through Letters, " "Getting a Job, " "Communication Through Reports, " "Oral Communication, " and "Managing Communication Activities. "   In addition to the comprehensive coverage of topics, the text provides detailed pedagogical devices, such as cases, discussion questions, and problems for writing.

365   Himstreet, William C. , and Baty, Wayne M.   Plaid for Busi-
          ness Communication.   Homewood, Ill.: Learning System
          Company, 1976.   163p.
"Plaid" stands for "Programmed Learning Aid. "   This book is for "people in business who have not taken business communication, " students preparing for competency tests in communication, and participants in in-service training programs.   The authors also suggest that it could be used as a supplement to a standard business communications text.   Each of the 20 subject divisions is organized by "frames. "   The user reads a frame, answers the questions at the end of the frame, and finds the answers on the next page.   Topics covered include communication theory, writing for various purposes, organizing information, résumés, speaking, and dictating.   Following the 20 topic divisions are four sets of review examinations, covering the subjects in the divisions.   The answers follow the examinations.

366   Himstreet, William C. ; Porter, Leonard J. ; and Maxwell,
          Gerald W.   Business English in Communications.   3rd
          ed.   Englewood Cliffs, N. J.: Prentice-Hall, 1975.

Index.   438p.

This textbook is geared to students in business administration who
need to strengthen their skills in communication--spoken and written.
The textbook is divided into three broad divisions: English fundamentals, including grammar and usage; forms of business correspondence, with writing exercises; and speaking and listening skills.
Letters with critiques are presented as well as writing exercises
and discussion questions.   Five appendixes contain advice on spelling, irregular verbs, salutations, proofreader's marks, manuscript
preparation.

367 Hirschhorn, Howard H.   Technical & Scientific Reader in Eng-
         lish.   N. p.: Regents, 1970. 203p.
This reader is designed "especially for Spanish-speaking students,
scientists, and technicians who are preparing to take courses in
American schools or under instructors abroad." The readings are
taken from such sources as engineering encyclopedias, a data-
processing glossary, a manual for jet engines, a refrigeration guide,
a primer of navigation, and a business math book.   Technical or
specialized words are defined in Spanish in the margin of the page
on which they appear.   The readings are all in English.   Each is
followed by three exercises: comprehension questions, a vocabulary
quiz, and a conversation exercise that asks the student to discuss
or explain a concept covered in the reading.   Exercise answers are
in the back of the book.

368 Hirschhorn, Howard H.   Writing for Science, Industry and Tech-
         nology.   New York: Van Nostrand, 1980.   Index.   265p.
This textbook is designed "for courses in scientific and technical
writing" and focuses on "practical ways to write the specialized documentation needed for today's industry, science, research, and technical management."   The 21 chapters are divided into three parts.
Part I covers the definition of technical writing, audience analysis,
and sources of scientific and technical information.   Part II covers
the elements of discourse (e.g., narration, description, argumentation), illustrations, and technical abbreviations and symbols.   Part
III covers outlines, letters and memos, technical instructions and
manuals, reports, articles, abstracts and annotated bibliographic
references, oral reports, revision, and documentation.   All chapters
contain writing projects or study questions.   The three appendixes
include an annotated list of information centers for scientific and
technical writers, a method for collecting and retrieving information,
and a metric conversion table.

369 Hodgson, Richard S.   The Dartnell Direct Mail and Mail Order
         Handbook.   2nd ed.   Chicago: Dartnell, 1974.   Index.
         Bib.   1,575p.
This massive handbook covers almost every aspect of direct-mail
advertising--from general principles to layout to addressing envelopes.
The author points out that there are few hard-and-fast rules or limitations to preparing direct mail, "except those imposed by postal
regulations, personal ethics, good taste ... and, of course, the
available budget."   In addition to providing an overview of the medium

of direct mail, this book of 49 chapters contains advice for planning, writing, organizing, formatting, and mailing. In addition, printing and postal regulations and terms are covered. The book also details cost analysis for ad campaigns and special equipment to use in direct mail-- including computers. The appendix of 272 pages consists primarily of tables.

370   Holscher, Harry H.   How to Organize and Write a Technical
          Report.   2nd ed.   Paterson, N. J. :   Littlefield, Adams,
          1965.   Index.   176p.
This book is designed as a brief writing manual for use with a "good English handbook, a dictionary, a thesaurus, and a grammar style man- ual. " The eight chapters cover such topics as the need for training in technical-report writing; differences between writing and speaking and the principles of reader awareness, organization, emphasis, and what management needs; the steps to preparing a report--need, outline, draft, revision, editing, charts and tables; physical format of a report; a re- vision checklist; and basic mechanical presentation in format and gram- mar. Chapter 7, "Some Interesting Writing, " gives samples of a tech- nical report, a scientific newspaper article, and a trade journal article. Chapter 8, "Appendix, " contains a table of abbreviations, list of over- used words, list of recommended words, table of copy-reading and proofreading symbols, and a list of often misspelled words.

371   Hookey, E. M.   Guide to the Preparation of Training Materials.
          Washington, D. C. :   Government Printing Office, 1962.

372   Hoover, Hardy.   Essentials for the Technical Writer.   New
          York: Wiley, 1970.   Index.   216p.
This book is specifically addressed to students and professionals in scientific and technical fields.   The five chapters cover planning and outlining; effective sentence construction; paragraph development; the report, structure, and style; and formats.   The 54-page appendix provides answers to the many exercises, definitions of types of tech- nical writing, and a vocabulary-building list of words for the student.

373   Horten, H. E.   Commercial Correspondence in Four Languages.
          New York: Hart, 1970.   316p.
This book is especially designed for exporters and importers, pro- viding a four-way guide to basic correspondence in English, French, German, and Spanish.   The book is set in four columns across two pages, giving the four-language versions of phrases, sentences, or words for such topics as letter endings, requests for special invoic- ing, and complaints.   The book provides a glossary of 1, 000 export- import terms in the four languages.

374   Hotchkiss, G. B.   Business Correspondence.   New York:
          Alexander Hamilton Institute, 1910.

375   Hotchkiss, George Burton, and Drew, Celia Anne.   New Business
          English.   New York: American Book, 1932.   Index.   394p.
This textbook, originally published in 1916 under the title Business English: Its Principles and Practice, covers most of the aspects of

business letters, including sections on advertising copy, reports, and legal points in letters.   The 25 chapters begin with general principles of "good business English"; the next eight chapters review language and writing.   Chapters 10-11 cover the format of letters.   The next 11 chapters deal with types of letters.   The last three chapters cover other forms of business writing (e.g., advertising copy and reports).   The chapter "Business Narrative" presents a series of letters about a grocery business.   This chapter functions as a case study and shows the resulting correspondence related to the case. The appendix contains reference sections, including legal points, abbreviations, antonyms and homonyms, glossary of business terms, proofreader's marks, syllabication, a true-false test for students, and a list of commonly misspelled words.   The book includes revision and writing exercises.

376   Hotchkiss, George Burton, and Kilduff, Edward Jones.   Advanced Business Correspondence.   4th ed.   New York: Harper and Brothers, 1947.   Index.   571p.
This text was originally published in 1921, written "with the assistance of J. Harold Janis."   It "presupposes ... familiarity with the requirements of good English technique" and "is intended primarily for classes of university grade."   The first three chapters review general letter-writing strategy, reader adaptation, and the essential qualities of successful letters.   Chapters 4-5 deal with the mechanics of good letters and prewriting.   Chapters 6-15 cover various types of letters: sales, adjustment, collection, credit, promotional, and routine.   Chapter 16 discusses developing mailing lists and tests, and Chapter 17 covers replies to sales inquiries.   Chapter 18 treats the use of argumentation in business letters; Chapter 19 covers "personal letters" in business; and Chapter 20 discusses how to improve and control business correspondence.   Chapter 21, the last chapter, covers business reports.   The appendix discusses the legal aspects of letters.

377   Hotchkiss, George B., and Kilduff, Edward J.   Handbook of Business English.   3rd ed.   New York: Harper and Brothers, 1945.   Index.   308p.
Originally published in 1914, this book is one of the early handbooks and is geared to people who write or dictate business messages. Six of the ten chapters are concerned with rhetoric and grammar: words, sentences, and paragraphs.   Chapters 2, 8, and 10 deal with business letter, its form and content, and business reports.

378   Houp, Kenneth W., and Blickle, Margaret D.   Reports for Science and Industry.   2nd ed.   New York: Holt, Rinehart and Winston, 1973.

379   Houp, Kenneth W., and Pearsall, Thomas E.   Reporting Technical Information.   4th ed.   Encino, Calif.:   Glencoe, 1980.   Index.   Bib.   547p.
This textbook is designed primarily for students in engineering and technology and professionals.   The 19 chapters are divided into three parts.   Part I serves as an introduction to technical-report

writing, covering investigating, planning, organizing, writing, and
revising.   There are also chapters on audience analysis and rhetorical modes.   Part II deals with report components: prose elements,
mechanical elements, and graphic elements.   Part III, "Applications,"
deals with correspondence, instructions, proposals, progress and
feasibility reports, etc.   Part IV is a handbook section.   The four
appendixes include a student report, science and engineering reference books, a brief history of measurement systems, and a selected
bibliography.   The exercises ask students to write and discuss problems and cases.

380   How to Collect Money by Mail.   Chicago: System Company,
            1913.

381   Howard, C. Jeriel, and Gill, Donald A.   Desk Copy: Modern
            Business Communication.   San Francisco, Calif.:   Can-
            field, 1971.

382   Howard, C. Jeriel; Tracz, Richard Francis; and Thomas, Cora-
            mae.   Contact: A Textbook in Applied Communications.
            3rd ed.   Englewood Cliffs, N.J.:   Prentice-Hall, 1979.
            Index.   295p.
This text of 17 chapters covers the needs of vocational students for
writing and speaking.   Chapters 1-5 review the importance of communication, types of letters, and general business-letter-writing
strategy.   Chapter 6 discusses the employment interview--how to
act, talk, dress, etc.; Chapter 7, the preparation of memos and
short reports; and Chapter 8, serving on a committee.   Chapters
9-10 cover basic logic.   Chapters 11-14 treat gathering information
from professional meetings, libraries, etc., and note taking and
reading.   Chapter 15 is on outlining; Chapter 16, oral report; and
Chapter 17, the formal report.   A 29-page grammar, usage, and
mechanics handbook is included.

383   Howell, Almonte C.   A Handbook of English in Engineering
            Usage.   2nd ed.   New York: Wiley, 1940.   Index.   Bib.
            433p.
This book is intended primarily for practicing engineers.   Chapters
1, 8, 9, and 10 deal with technical writing (letters, reports, magazine articles).   Chapters 2-7 cover usage, grammar, and paragraphing.   Three appendixes include the presentation of statistics, interviews and questionnaires, and editorials and book reviews.

384   Howell, Almonte C.   Military Correspondence and Reports.
            New York: McGraw-Hill, 1943.   Index.   190p.
This book is aimed at those military personnel who needed a writing
guide and a "convenient digest of rules and regulations governing
[military] report and letter writing."   Chapter 1, "Military Correspondence," covers both form and style.   "The writer should ... not
use the first personal pronoun ... not even the second person should
be used."   Chapter 2, "Military Reports," describes the layout and
typical content of reports.   Chapter 3, "General and Special Orders,"
is on the military form, language, and tone of orders.   Chapter 4

covers "Bulletins, Circulars, and Memoranda," and Chapter 5, "Field Orders, Reports, and Messages." These chapters treat, in particular, conciseness, accuracy, and specificity in these formats as well as form requirements. The six appendixes make up over half of the book. Appendix A (37 pages) is a handbook of usage, grammar, and abbreviations. Appendixes B and C are digests of civilian and naval correspondence forms. Appendix D is a bibliography of army regulations and style handbooks. Appendix E is an outline of a brief course in military writing. Appendix F contains outlines and examples of army reports.

385 Huber, Jack T. Report Writing in Psychology and Psychiatry. New York: Harper and Row, 1961. Index. Bib. 114p. In the first of the eight chapters, this book blends reader analysis and the report's objective with the special needs of psychology and psychiatry. Chapter 2, "Formulating the Case," covers establishing a perspective for the report, and the methods of organization (chronological, topical, and both). Chapter 3, "Outlines for Reports in Special Areas," presents general guidelines for outlining and gives typical outlines for reports on intelligence, neurological studies, and vocational and psychiatric cases, among others. Chapter 4, "Therapy Progress Notes," discusses writing and assesses the need for notes. Chapter 5, "Some Notes on What to Write," covers analysis and interpretation in reports. Chapter 6, "How to Put It in Writing," explains the principles of clear style, use of jargon, and some mechanical details. Chapter 7 is on "The Dilemma of Confidentiality." Chapter 8 has "Some Examples of Report Writing."

386 Hunsinger, Marjorie. Business Correspondence for Colleges. New York: Gregg/McGraw-Hill, 1960. Index. 250p. According to the author of this text-workbook, students are led "by the hand ... through the principles and processes of letter writing so that learning becomes easy and almost fun." The book is divided into six parts, which are further subdivided into 21 topics. Each section of this 8½" x 11" book contains tear-out worksheets that the students use to write the assignments. Part I is an overview of the good business letter. Parts II-III cover the format of letters and the principles of good letter writing. Part IV, "Writing the Letter," deals with planning, outlining, and style. Part V, "Common Types of Business Letters," covers order, inquiry, goodwill, sales, adjustment, credit, application, and routine letters. Part VI, "Other Forms of Business Writing," provides a five-page discussion of memos and reports. The reference section is devoted to punctuation and grammar.

387 Hunsinger, Marjorie, and Clarke, Peter B. Modern Business Correspondence: A Text-Workbook for Colleges. 3rd ed. New York: Gregg/McGraw-Hill, 1972. Index. 252p. This text-workbook has 22 units divided among six parts: "Letters in Business," "Mechanics of Good Letter Writing," "Principles of Good Letter Writing," "Writing the Letter," "Common Types of Business Letters," and "Other Forms of Business Writing." Under Part

II, "Mechanics, " are included letter styles, usage, and stationery
styles.   Each unit has an explanation of the topic along with illustrations.   Assignments appear throughout, geared to the workbook
section, which is labeled by unit number.   The reference section is
a brief grammar review.   The workbook section contains three "Language Checkup" sheets to practice grammar problems.   Workbook
exercises include rewriting letters, doing drafts in answer to problems, practice in organization, and revising excerpts for clarity and
goodwill.

388 Hunsinger, Marjorie, and McComas, Donna C.   Modern Busi-
            ness Correspondence: A Text-Workbook for Colleges.
            4th ed.   New York: Gregg/McGraw-Hill, 1979.   Index.
            206p.
This edition of this spiral-bound textbook is the first with McComas
as coauthor.   The 21 units are divided into four parts.   Part I, "The
Appearance of Business Letters, " covers stationery, format, and
grammar.   Part II, "Principles of Good Letter Writing, " discusses
word choice, sentences and paragraphs, tone, and goodwill.   Part
III, "Writing Effective Business Letters" deals with types of letters,
such as sales, inquiries, collection, and credit.   Part IV, "Principles of Report Writing, " is a one-unit treatment of format.   Worksheets for each unit appear at the back of the book.   A reference
section covers punctuation, parts of speech, abbreviations, and numbers.   The index is in the beginning of the book.

389 Hunter, Laura Grace.   The Language of Audit Reports.   U. S.
            General Accounting Office.   Washington, D. C. :   U. S.
            Government Printing Office, 1957.

390 Hutchinson, Helene D. , ed.   Horizons: Readings and Commu-
            nication Activities for Vocational-Technical Students.
            Beverly Hills, Calif. :   Glencoe, 1975.   324p.
According to the author, the book "is a multi-genre collection of
readings for the vocational-technical student. "   Each reading is followed by exercises (speaking, writing, and listening).   "Because the
vocational-technical student is a person first and a technician second, this text seeks to broaden his view and extend his horizon both
personally and professionally. "   The 81 items include essays, short
stories, poetry, tests, questionnaires, directions, photographs, and
drawings.

391 Institute in Technical and Industrial Communication.   Annual
            Proceedings: 1958--Present.   Fort Collins: Colorado
            State University Press.   Pages vary.
This collection of papers presented at the ITIC meetings is a reference tool for educators, scholars, and practitioners.   Each institute
has a theme, such as "Quality Production of Technical Publications
with a Declining Budget and Computer-Aided Instruction" (1971).
Contributors from business, industry, and academe are represented
in the articles, which reflect trends and methodology.

392 International Trade.   English Language Service.   New York:
            Collier/Macmillan, 1966.

393 Ironman, Ralph. Writing the Executive Report. New York:
      Funk and Wagnalls, 1966. 145p.
According to the author, this guide is for "those engaged in science,
technology, and management." The examples used and some of the
stylistic advice grows out of British usage. After a brief introduction, in which the author points out that some parts of the eight
chapters are aimed particularly at European writers, Ironman defines
and classifies report types. "The Ideal Report" covers planning,
writing, and general advice. A six-page chapter on readability deals
with transition, sentence length, wordiness, and pronoun use. Following is "Some Literary Difficulties," covering style problems of
grammar, punctuation, spelling, and other mechanics. The final
three chapters review professional technical writing, technical illustration, reproduction, and the function of a technical editor. Three
appendixes provide proofreading symbols, a commentary on basic
reference material, and examples of "weekly news sheets."

394 Jackson, Clyde W. Verbal Information Systems: A Compre-
        hensive Guide to Writing Manuals. Cleveland: Associa-
        tion for Systems Management, 1974. 74p.
The purpose of this book is to help the writer of manuals, which the
author defines as "verbal information systems." He uses this term
because manuals are an "operating and management tool" that use
words to function. Within the 18 brief chapters, the author explains
the various roles of manuals in organizations and defines standards
for good manuals. He also covers writing style and format. Topics
include the use of playscript procedure for constructing manuals, relationships of writing departments to other departments and technicians, and creating an effective package for the manual. Numerous
parts of manuals are provided as illustrations throughout this book.

395 Jacobi, Ernest. Work at Writing: A Workbook for Writing at
        Work. Rochelle Park, N. J.: Hayden, 1980.

396 Jacobi, Ernest. Writing at Work: Dos, Don'ts and How Tos.
        Rochelle Park, N. J.: Hayden, 1976. Index. 198p.
An outgrowth of letters from a father to his son in college, this book
is designed for professionals and people in business. The conversational tone is maintained through colloquial chapter titles and scat-
tered anecdotes. The 46 short chapters provide numerous tips, including discussions of indexing, press releases, and how to be or
use a good editor. The last third of the book covers recurring
grammar problems, information on printing and other means of reproduction, and psychology in writing. The book ends with exercises
and three appendixes, covering visuals, inductive/deductive writing,
and a proofreading guide.

397 Jaggi, Willy. The Manuscript. 6th ed. Basel, Switzerland:
        Karger, 1969. 48p.
This guide is for the preparation of manuscripts and bibliographies,
especially in the medical and veterinary sciences. The seven sections discuss general principles; manuscript format for articles,
books, and offset printing; illustrations; corrections; and bibliography

format. The section on abbreviations includes a list of general terms
and geographical locations and a 19-page list of international journals and yearbooks for medicine and veterinary science. A "Units"
section provides a seven-page list of symbols, formulas, and measurements in English, French, and German. Headings, proofreader's
marks, and marginal notations are printed in red.

398 Janis, J. Harold. Business Communication Reader. New
        York: Harper and Brothers, 1958. Index. 369p.
This anthology of articles is divided into ten chapters, each of which
contains at least four articles and is followed by an exercise section
of discussion questions, writing assignments, and questions on the
readings. Among the topics covered are "Moral Values in Business, "
"The Businessman in American Literature, " and "Preparing for Business Leadership. "

399 Janis, J. Harold. The Business Research Paper. Business
        Communication Series. New York: Hobbs, Dorman,
        1967. Index. Bib. 100p.
This guide is aimed at graduate students preparing theses or dissertations in business fields. Chapters include information on documentation, research methods, format, tables, and a "Checklist for Writers. " The appendix is a sample of papers from Master's theses
from NYU. A short bibliography on research papers is included.

400 Janis, J. Harold. Writing and Communicating in Business.
        3rd ed. New York: Macmillan, 1978. Index. Bib.
        548p.
This textbook aims to teach not only the "what" of business communication but also the "why. " The first two of the 16 chapters, therefore, are devoted to the theory of communication in organizations.
Chapter 3 presents the "Language of Business, " and Chapter 4, the
general principles of business-letter writing. Chapters 5-7 are devoted to types of letters. Chapters 8-15 cover formats and strategies, including organization, persuasion, reports, graphs, sales and
goodwill letters, and handling controversial issues. Chapter 16 is
an 18-page treatment of speech communication. Following the chapters are a list of reference books and a style manual. Chapters end
in exercises and cases are included.

401 Janis, J. Harold, and Dressner, Howard B. Business Writing.
        2nd ed. College Outline Series. New York: Barnes and
        Noble, 1972. Index. 402p.
This book, first published in 1956 under the title Business English,
deals with general writing principles and specific types of writing.
Part I, "Written Communication in Business, " discusses the scope
and function of written communication and describes models of the
communication process. Part II, "Business Letters, " covers the
format, reader analysis, effective style, organization, and types of
letters. Part III, "Business Reports, " is on planning and researching, organizing data, and graphics. Part IV is a "Handbook of
Grammar and Usage" reviewing grammatical terms and rules. Exercises cover the text material, ask for revision, or correct gram-

mar. Answers to the grammar exercises are given in the back of
the book.

402 Janner, Greville (Mitchell, Ewan, pseud.). The Businessman's
       Guide to Letter-Writing and to the Law on Letters. 2nd
       ed. London: Business Books, 1977. Index. 222p.
This book of 72 chapters is divided into eight parts. The first four
parts treat letter form style and grammar, tact and tactics, and
types of letters (e.g., sales, congratulations, condolence, and collection). Part V covers office systems, duplicating, stationery, etc.
The final three parts cover letters and the law: definitions of legal
points, types of legal letters, and litigation involving letters. The
informal style throughout is reflected in cartoons about letter writing. The three appendixes cover postal information, abbreviations,
and foreign usage.

403 Janner, Greville (Mitchell, Ewan, pseud.). Letters of the
       Law: The Businessman's Encyclopedia of Draft Letters
       with Legal Implications. New York: International Publi-
       cations Service, 1970.

404 Jeffares, A. Norman, and Davies, M. Bryn. The Scientific
       Background: A Prose Anthology. London: Pitman, 1958.
       306p.
This collection of 100 essays and excerpts from writings on science,
writing, and the humanities is designed to solve what the authors
call "the most pressing educational problem of the moment": that is,
"the broad divergence brought about by specialization between the two
main branches of knowledge, the sciences and the humanities. " The
author concludes that "good English scientific prose is good English
prose. " The collection begins with an excerpt from Thomas Sprat's
History of the Royal Society, titled "The Prose Style Sought by the
Royal Society. " Selections include writing by scientists, such as
Darwin and Newton, and by literary figures, such as Samuel Johnson
and Keats. Also included are selections on writing itself (e.g., H.
W. Fowler's "Preposition at End" and Ben Franklin's "Advice About
Writing. "). Following the selections are discussion questions.

405 Johnson, Thomas P. Analytical Writing: A Handbook for Busi-
       ness and Technical Writers. New York: Harper and
       Row, 1966. Index. Bib. 245p.
This book, aimed at writers in business and technology, advocates
an "analytical" approach. That is, the author assumes that when a
writer in business fails to communicate with a reader, "he has not
bothered to analyze or interpret the information he has presented"
to that reader--he has taken a "catalogical" approach to writing.
The 16 chapters are divided into five parts. Part I, "What Business
and Technical Writing Is All About, " is composed of a single chapter
that argues that good writing is based on logical presentation, not
grammar. Part II (seven chapters), "Five Ways to Test Business
and Technical Writing, " discusses aspects of poor writing: lack of
detail, sentence variety, and subordination, and the use of abstract
nouns, passive verbs, and impersonal style. Part II concludes with

ten case histories of pieces of business and technical writing.   Part
III, "The Logical Flow of Ideas (Putting Sentences Together)," covers
paragraph structure, orienting the reader, avoiding technical density,
and transitional devices.   Part IV, "The Structure of Business and
Technical Writing," deals with organizational patterns, openings, and
organizing long formal reports.   It also provides eight case histories.
Part V, "Metaphor in Business and Technical Writing," is a ten-page
chapter on the use of metaphor.

406   Jones, Everitt L., and Durham, Philip.   Readings in Science
        and Engineering.   New York: Holt, Rinehart and Winston,
        1961.   364p.
This book is a collection of scientific writings.   Besides serving as
examples of scientific definition, exposition, narration, and argument,
these readings also indirectly instruct the student of science in how
to organize scientific data or material and present it effectively.
One section in the book, "About Writing," has seven units as the
nature, methods, and styles of scientific and technical writing.

407   Jones, J. Stanley.   English for the Business Student.   1963;
        rpt. London: Evans Brothers, 1974.   212p.
This "course in English" was written "to meet the needs of students
following Commercial Courses in Secondary Schools, Technical and
Commercial Colleges and other establishments of Further Education
and in Evening Commercial Classes."   The author says that the
course should provide preparation for major examinations of such
groups as the Royal Society of Arts, the London Chamber of Commerce, and the Institute of Bankers.   The 15 chapters cover such
topics as word choice, combination of sentences, clarity, grammar,
paragraphs, metaphor, business correspondence, and reports.   All
chapters have exercises and problems for writing.   Chapter 6 discusses using reference books and provides exercises.

408   Jones, W. Paul.   Writing Scientific Papers and Reports.   7th
        ed.   Dubuque, Iowa: Brown, 1976.   Index.   366p.
This book, originally published in 1946, was designed "for a course
in the writing of scientific papers and reports in the scientific and
engineering curricula of Iowa State University."   It remains directed
to science and engineering students at the upper or graduate levels.
The first of 20 chapters is titled "The Method of Science" and places
writing within the context of the scientific method of research.   Following the three introductory chapters are nine chapters describing
methods and types of scientific and technical writing, such as definitions, analyses, sets of directions, and popular-science articles.
Then come four chapters dealing with the writing of reports and
proposals.   The last three handbook-like chapters deal with documentation, grammar, punctuation, and style.   Chapters end in writing exercises.

409   Jones, Walter P., and Johnson, Quentin, eds.   Essays on
        Thinking and Writing in Science, Engineering and Busi-
        ness.   3rd ed.   Dubuque, Iowa: Brown, 1963.   297p.
This anthology was "assembled primarily because of the value [the

essays] have for students of science, engineering, and business who
are learning to write. " The 37 selections are organized into six
sections: "Why Study English?, " "The Scientific Method, " "Kinds of
Scientific Writing, " "Good English and Bad English, " "Readability
Formulas, " and "Information and Communication. " Comprehension
questions follow each essay.

410   Joplin, Bruce, and Pattillo, James W.   Effective Accounting
          Reports.   Englewood Cliffs, N. J. :   Prentice-Hall, 1969.
          Index.   251p.
This book was written "for the practicing accountant who wants to
prepare better--more readable and more effective--reports. " The
book covers determining the reader's needs, designing for effective
physical layout, writing in a clear style, and using visual aids.   In
addition, the book, in 11 chapters, covers topics of special interest
to the accountant, such as designing a reporting system, creating
"exception" reports, control of reports, and computer-generated reports.   The final chapter is a series of typical accountant's reports.
The book is in $8\frac{1}{2}$ " x 11", size.

411   Jordan, Edwin Pratt, and Shebard, W. C.   Medical Writing.
          Philadelphia: Saunders, 1958.

412   Jordan, Richard C. , and Edwards, Marion J.   Aids to Tech-
          nical Writing.   Bulletin No. 21.   Minneapolis: University
          of Minnesota Engineering Experiment Station, May 15,
          1944.   Index.   Bib.   117p.
This bulletin is aimed at presenting a guide to style of presentation,
general format, and illustrations for technical writing.   The focus is
on mechanical engineering.   The 16 chapters include "Style in Non-
Letterpress Publications, " "Bibliography Forms, " "Numbers, " "Tables, " "Photography in Technical Publications, " and "Graphic Symbols. "   There are many illustrations of the formats covered and
definitions of terms.   The authors stress that the guide is not for
a writer, but for an engineer who needs to write technical publications.

413   Jordan, Stello, ed.   Handbook of Technical Writing Practices.
          2 vols.   New York: Wiley-Interscience, 1971.   Index.
          Bib.   1, 374p.
This two-volume publication is designed to help professional technical
writers with every possible publication task.   In fact, it is a collection of articles written by leaders in the field of technical writing on
subjects ranging from writing reports and parts catalogs to managing
technical-publication departments.   The two volumes are divided into
four parts.   Part I, "Documents and Publications, " is the largest
and comprises the first volume entirely.   Part II, "Supporting Services for Technical Writing, " discusses illustrating, production, and
editing.   Part III, "Management of Technical Writing, " covers numerous aspects of the technical-publication department, including the role
of technical writing activities and supervision.   Part IV, "Guides and
References, " serves as a literature review for the field of technical
writing up to 1970.   This work is a comprehensive source of infor-

mation on writing, forms, and formats for the technical writing teacher.   Each volume has its own index.

414   Kakonis, Thomas E., and Scally, John.   Writing in an Age of
        Technology.   New York: Macmillan, 1978.   Index.   Bib.
        355p.
This text is designed for freshman composition courses for students in technological fields.   The theme in the text is that effective communication is "crucial to survival and success in the world of economic realities."   Units I and II deal with sentences, paragraphs, and modes of discourse.   Unit III, "Writing for the 'Real' World," offers 17 essays and articles dealing with the relationships between humanistic and technological values.   Selections range from articles from Time and The Wall Street Journal to essays from The Futurist and Technology and Culture.   Each selection is followed by questions on writer's purpose, rhetorical devices, organization, and effectiveness of technique.   Unit V, "Job-Related Writing," deals with short reports, proposals, and specifications, instructions, and résumés. Included in the text is a "Glossary of Frequent Errors."

415   Kantrovich, Jerald M.   Rifle the Deck: Technical-Vocational
        Writing for High School.   Chicago: Stack the Deck, 1978.

416   Kapp, Reginald O.   The Presentation of Technical Information.
        New York: Macmillan, 1957.   Index.   147p.
Long considered a classic, this book is based on four lectures the author gave at University College, London.   Originally published in 1948, most of the advice remains timeless.   Among the topics covered by the 15 chapters are "Functional English," "Functional and Imaginative Literature Compared," "Making It Easy to Understand," and "Metaphor."

417   Kearney, Paul W.   Business Letters Made Easy: How to Win
        Success in Business Through Effective Correspondence.
        Boston: Books, 1939.   297p.
This book is geared to the business writer on the job.   The first three chapters deal with the essential qualities of letters: simplicity, the "you attitude," and conciseness.   The book emphasizes that letter writing is talking on paper; hence, naturalness is the quality to be cultivated.   Chapter 3 covers punctuation and outmoded expressions.   Chapters 4-9 deal with the format of letters, organization, and types of correspondence.   Model letters are provided. Chapter 10 gives advice to those who wish to do consulting on letter writing in industry.

418   Keithley, Erwin M., and Schreiner, Philip J.   A Manual of
        Style for the Preparation of Papers and Reports.   2nd
        ed.   Cincinnati: South-Western, 1971.   Index.   Bib.
        81p.
This style guide covers the preparation of reports for college, business, industry, or science.   Topics include "Arranging, Editing, Typing, and Binding," "The Body of the Report--Documentation," "The Bibliography."   The appendix is a model report.   The last

page is a sample guide sheet for use in typing--to be placed behind
the sheet being typed.

419   Keithley, Erwin M. , and Thompson, Margaret H.   English for
         Modern Business.   3rd ed.   Homewood, Ill. :   Irwin,
         1977.   Index.   418p.
This book is written for students who need a refresher course in
grammar and usage.   The authors suggest that students take this
refresher course before they take a course in writing.   The introduction contains an 18-week syllabus, a ten-week term syllabus, and
suggestions for the teacher on how to handle assignments and give
exams.   The teacher's edition of the text contains eight exams with
answers.   There are 12 sections, each with several parts.   Topics
include sentence structure, punctuation, parts of speech, abbreviations, numbers, and vocabulary skill.   Each part has exercises on
tear-out sheets.   Four appendixes cover standard abbreviations, prepositions in special combinations, grammatical terms, and spelling
500 difficult words.   All examples use business context.

420   Kelly, R. A.   The Use of English for Technical Students.   2nd
         ed.   London: Harrap, 1970.   189p.
This British text offers extensive exercises in technical and scientific writing to meet the "demand of students at Ordinary National
Certificate level and at some stages of the City and Guilds courses. "
The eight chapters cover "Sentences, " "Paragraphs, " "Description
and Explanation, " "Special Types of Exposition" (instructions, letters,
reports), "Comprehension and Interpretation" (reading), "Note-Taking, "
and "Summarizing. "   Chapter 8, "Technical Vocabulary, " covers
spelling rules, special terms, word choice, and specialized spelling
lists for trades.   All chapters have extensive sentence exercises and
writing assignments.

421   Kennedy, W. W. , and Bridges, T. B.   Effective English and
         Letter Writing.   Battle Creek, Mich. :   Ellis, 1918.

422   Kent, Sherman.   Writing History.   New York: Appleton-
         Century-Crofts, 1967.   Index.   Bib.   143p.
Aimed at both undergraduate and graduate students, this book helps
the student write historical essays.   It begins with an overview of
the purpose and method of historical writing.   Parts I and II cover
finding and limiting a topic.   Part III covers scholarly research,
and Part IV, methods of organization.   "Historical writing must
have ... a continuous flow of clearly stated ideas.   In this respect
it differs most essentially from, say, the writing of belles lettres.... "
Parts V and VI cover effective style.   Part VII deals with making
an index.   There is an 11-page bibliography providing sources for
the history student.

423   Kenyon, H. S.   Spanish Commercial Correspondence.   Ann Ar-
         bor, Mich. :   Wahr, 1907.

424   Kimball, G. S.   Business English.   Indianapolis: Bobbs-
         Merrill, 1908.

425 King, F. W. , and Cree, D. Ann. Modern English Business
     Letters. New York: Longman, 1975.

426 King, Lester S. , and Roland, Charles G. Scientific Writing.
          Chicago: American Medical Association, 1968. 133p.
This book is a collection of articles written by the authors (two doctors) after they gave courses in medical writing. It is aimed at
physicians who engage in medical writing. Topics include parts of
speech, passive voice, jargon, clichés, outlining, openings, abstracts,
case reports, and rewriting. The last three chapters analyze the
styles of Macaulay, Carlyle, and Samuel Johnson. One chapter includes a list of "Writers' Reference Books."

427 Kirby, Richard S. The Elements of Specification Writing: A
          Textbook for Students in Civil Engineering. New York:
          Wiley, 1935. Index. 168p.
This book, originally published in 1913, is a legal as well as a
writing guide for the preparation of construction specifications for
students and practicing engineers. Chapter 1 begins with a definition of terms and the financial arrangements for construction specifications. Chapter 2 defines and describes the advertisements for
bidders. Chapter 3 covers the necessary elements of proposals.
Chapters 4 and 5 cover legal questions. Chapter 6, "Specifications
and Their Composition," discusses the characteristics of good writing in specifications. Chapters 7-13 focus on specific clauses in
specifications, and Chapter 14 presents case problems for the student.

428 Kirkman, John, ed. Teaching Communication Skills to Engi-
          neers and Scientists in Higher Education. Cardiff: Univer-
          sity of Wales Institute of Science and Technology, 1978.

429 Kirkpatrick, Thomas Winfred, and Breese, M. H. Better
          English for Technical Authors or Call a Spade a Spade.
          New York: Macmillan, 1961. Index. 122p.
This book grew out of the authors' experience teaching postgraduate
students at the Imperial College of Tropical Agriculture. The examples used are from agricultural contexts, and the book is aimed
at agriculturists who must write reports, theses, papers, and books.
Some of the 30 chapters are one page long; however, the main sections are Part I, "Mainly on Style," covering conciseness, word
choice, ambiguity, etc.; Part II, "Mainly on Grammar," on such
topics as subject/verb agreement, tenses, and number; and Part III
(untitled), on grammar and other "Sundry Subjects." The chapters
have a serious tone, but many illustrations are humorous.

430 Klopfer, Walter G. The Psychological Report: Use and Com-
          munication of Psychological Findings. New York: Grune
          and Stratton, 1960. Index. Bib. 146p.
This book is designed for students and professionals in psychology
and psychiatry. The first of nine chapters provides background on
the purpose of the psychological report. The second chapter, "Focus," suggests several psychological points of view for reports:

treatment planning, administrative planning, and prediction of behavior.   Chapter 3 deals with writing style, and Chapters 4-5 suggest
patterns of organization.   Chapter 6 reviews "language" of reports
and appropriate ways for handling technical terminology.   Chapters
7-8 cover the influence of several kinds of bias in the report.   Chapter 9 presents a case study of a 14-year-old boy and the conclusion
that could be drawn in a report.   The appendix contains sample
psychological reports.

431   Knapper, Arno, and Newcomb, Loda.   Style Manual for Written
        Communication.   Columbus, Ohio: Grid, 1974.   Index.
        202p.
This handbook is a style manual prepared for writers whose objective
is "communication."   The 14 chapters cover letter styles (format,
signature lines, folding), report mechanics (footnotes, listing, format), grammar, punctuation, proofreading, table construction, abbreviations, usage, trite expressions, etc.   There is also a unit on
dividing words at the end of a line and an eight-page unit on idiomatic prepositions.

432   Kobe, Kenneth Albert.   Chemical Engineering Reports: How to
        Search the Literature and Prepare a Report.   4th ed.
        New York: Interscience, 1957.   Index.   Bib.   175p.
This book for students of chemical engineering and practicing engineers takes the form of a report.   It uses typewriter font on $8\frac{1}{2}$" x
11" pages, includes a letter of transmittal to the reader rather than
a preface, and generally follows the conventions of a formal report.
Among the topics covered are the research literature available to
chemical engineers, types of reports, business letters, principles
of effective writing, the mechanical preparation of reports, and bibliography preparation.   The appendix lists abbreviations and symbols
and includes the series of three articles, "Engineers Can Write Better," by H. J. Tichy.

433   Koch, Robert.   What You Should Know About Business Writing.
        Business Almanac Series No. 9.   Dobbs Ferry, N.Y.:
        Oceana, 1967.   Index.   91p.
This book is one in a series of short books designed to introduce the
reader to such topics as customer relations, direct mail, reducing
credit losses, etc.   This book has five chapters: "Business Writing Made Easy?," "You and Business Writing," "When and How to
Write (or not Write)," "Writing the 'Right' Business Message," and
"Was It a Profitable Message?"   Chapter 2 includes a checklist for
self-analysis of writing strengths and weaknesses.

434   Koelsche, Charles L., and Morgan, Ashley G., Jr.   Scientific
        Literacy in the Sixties.   Athens: University of Georgia,
        1964.   Bib.   38p.
This report is based on a study conducted to answer the question,
"What does a person need to know in order to be scientifically lit-
erate in the sixties?"   The researchers studied publications from a
number of scientific fields.   Included is a list of words "needed for
interpreting and understanding science-oriented articles."   The au-

thors also report what they consider the central scientific facts needed to answer their questions.

435 Koestler, Frances A. Creative Annual Reports: A Step-by-
         Step Guide. New York: National Public Relations Coun-
         cil of Health and Welfare Services, 1969. 71p.
This work is one of a series of how-to-do-it guides. The nine chapters deal with audiences for annual reports, organizational patterns
for reports, good openings, honesty in reporting, statistics, graphics, and distribution. The style is chatty--one professional to another. There is an emphasis on "tips, " such as using typewriter symbols to make charts (0 + X = $\overset{0}{X}$ a child). There are many illustrations taken from annual reports.

436 Kolin, Philip C. , and Kolin, Janeen L. Professional Writing
         for Nurses in Education, Practice, and Research. St.
         Louis: Mosby, 1980.

437 Konikow, Robert B. , and McElroy, Frank E. Communications
         for the Safety Professional. Chicago: National Safety
         Council, 1975. Index. 528p.
This manual is the result of the authors' ten years of planning and
gathering information in connection with the National Safety Council.
There are three major subject areas: "Communications Background"
covers nature and theory of communication and methods for improving communication. "Proven Ideas" includes material on meetings
and conferences, speeches, training, campaigns and contests, award
presentations, newsletters, reports, articles, and suggestion systems. "Graphic Communications" discusses presentation of information, emphasizing composition, arrangement and layout, using audiovisual media and techniques, and preparing effective presentations.
There are more than 170 illustrations.

438 Kramer, Edward. How to Punctuate a Business Letter. 2nd
         ed. New York: Pitman, 1960.

439 Krathwohl, David R. How to Prepare a Research Proposal.
         2nd ed. Syracuse, N. Y. : Syracuse University Book-
         store, 1977. Index. 112p.
This guide is for researchers in the behavioral sciences who are
seeking funds. The guide covers the principles of proposal writing.
Section I is "A Checklist for Research Proposals, " with page numbers keyed to the other sections. Section II, "General Comments on
the Preparation of Proposals, " covers purpose and scope of the proposal. Section III, "The Preparation of the Proposal, " reviews
problem, procedure, budget, data, personnel, and format. Section
IV, "Aids to Proposal Preparation, " covers reasons for failures to
get money, Federal and Foundation funding, particular kinds of proposals, and design of study. Section IV includes "Some Writing
Tips, " advising clear and concise language.

440 Krey, Isabelle A. , and Metzler, Bernadette V. Principles and
         Techniques of Effective Business Communication. New

York: Harcourt Brace Jovanovich, 1976. Index. 532p.
This text-workbook is divided into three sections: "Elements of
Effective Writing, " "Social-Business and Personal-Business Writing, "
and "Writing on the Job. " Within the 14 chapters are units on, for
example, "The Thank-You Letter, " "The Letter of Inquiry, " "Focus
on Language, " and "The Three C's of Credit. " Interspersed throughout the text are 71 skills activities, which present exercises or long
case problems. There is an answer key in the back of the book.
This text contains numerous exercises and assignments. Pages are
perforated so that students can turn in sheets. There are many illustrations of style in forms of address, closing, and tips on what
to include in types of letters. Wide margins contain italicized reminders or think questions, such as "Remember, a letter is an official record, " or "Would you make any changes in this letter?"
Also in the margins are definitions of words used in the text.

441 Kruse, Benedict, and Kruse, Bettijune. English for Business:
     Marketing. New York: McGraw-Hill, 1976. Index.
     120p.
The purpose of this book is to acquaint foreign students or foreign
business people with basic terminology and procedures in U. S. business offices. There are ten chapters covering such topics as "Telephone Sales, " "Purchasing, " "Credit, " "Inventory Control, " "Product
Management, " "Advertising, " "Market Research, " and "Management. "
Each chapter has a simple explanation of what someone who works in
these areas does in his or her job. For example, Mary works in
customer service, and Steve is an outside salesman. After the explanation are questions to establish that the student understands the
material. "Written Review" asks students to fill in sentences and
practice word choice. Finally, the chapters have a little dramati-
zation showing Steve and Mary in action on the job. The glossary
defines common business terms.

442 Laird, Dugan. Writing for Results: Principles and Practices.
     Reading, Mass.: Addison-Wesley, 1978. 266p.
This extensive revision of Business Writing Skills: A Workbook
(1970) provides more information on the processes of business writing, includes business exercises, and eliminates "the truly embar-
rassing male chauvinism" in the early book. Chapters 1-5 cover
general concepts of business writing, letter formats, organization
principles, sentence clarity, punctuation, friendly attitudes, and report writing. Included is a list of eight recommended books on report writing. Chapter 6, "The Chris Dawson 'In-Basket': A Simulation, " asks students to imagine that they are Chris Dawson, Manager for Electro Engineering. They are to handle correspondence for
all the items in the in-basket (letters, memos, reports). Writing
assignments, discussion questions, and exercises follow each of the
first five chapters.

443 Laird, Dugan, and Hayes, J. R. Level-Headed Letters. New
     York: Hayden, 1964. Index. 139p.
This book is designed for people in business who must write even
though they are afraid of and are unskilled at writing. Section I

aims to answer the questions used as chapter titles: "Why Do We
Talk Better Than We Write?," "Why Don't We Write the Way We
Talk?," and "Why Do We Write Letters?" Section II covers organization, logic and outlining, good letter-writing psychology and responding to the letters of others. The bulk of Section III is contained in Chapter 8, "How Much Grammar Do I Need?" The 72-
page chapter includes a pretest, posttest, and a programmed review
of grammar. Chapters 9 and 10 cover conciseness and the use of
a friendly attitude. Section IV contains Chapter 11 (a pep talk for
the reader) and the index.

444   Lamar, L.   Pattern and Purpose in Writing.   New York: Holt,
          1963.

445   Lamb, Marion M., and Hughes, Eugene H.   Business Letters,
          Memorandums, and Reports: A Basic Text in Business
          Communication.   New York: Harper and Row, 1967.
          Index.   556p.
This text is three-hole punched for insertion in a binder. The authors acknowledge the assistance of 98 people and organizations that
supplied writing samples for use in the discussions and in the exercises, which emphasize editing. Other assignments stress library
use. The 12 chapters cover effective writing of letters, memos,
and reports. Chapter 6, "Your Letters and the Law," discusses
business letters in relation to libel, fraud, and invasion of privacy.
Chapter 12, "Business Aspects of Business Writing," stresses the
handling of mail, classes of mail, dictation, and checking and sign-
ing letters. There is a "Supplementary Assignment Section" containing further material for student practice.

446   Lane, Nancy D., and Kammerer, Kathryn L.   Writer's Guide
          to Medical Journals.   Cambridge, Mass.:   Ballinger,
          1975.   Index.   Bib.   327p.
This guide is divided into five sections. The "Introduction" explains
how to use the book. The "Classified Subject Guide" lists the major
branches of medicine alphabetically. Under each classification, the
titles of journals devoted to that area are listed with page numbers
showing where they are discussed. The "Journal Publication Data
and Instructions" section lists each journal alphabetically, including
such information as editors' names, publication time lag, percentage
of submissions accepted, frequency of publication, and the journals'
instructions to authors up to ten pages in length. Following these
sections is a permuted title index and a bibliography of writing and
style manuals.

447   Lannon, John M.   Technical Writing.   Boston: Little, Brown,
          1979.   Index.   605p.
This textbook, "intended specifically for heterogeneous classes,"
provides numerous in-class and out-of-class exercises and writing
assignments. The 15 chapters are divided into three parts. Part
I covers the specific needs of technical writing and audience analysis. Part II covers rhetorical and organizational strategies, such
as summarizing, defining, classification and division, organizing,

researching, and using effective formats and visual aids.   Part III
treats specific applications, including letters, reports, descriptions,
processes, and oral reports.   Appendix A reviews grammar, usage,
and mechanics; Appendix B provides advice on the brainstorming
technique.

448   Larsen, Spencer A. , ed.   How to Improve Business Communi-
          cations as Told at the Business Communications Confer-
          ence, Wayne Univ. April 13, 1950.   Detroit: Wayne
          University Press, 1951.   221p.
This book is a collection of conference presentations made by academics and practitioners.   The topics include both very specific
items, like "Auditors Reports, " and large issues, like "Human Relations in Business Communications. "   The book is divided into 15
sections, among them "Creating Advertising Copy, " "Vitalizing Manu-
facturers' Publications, " "Improving the Day-to-Day Letters of Business, " "Legal Considerations of Business Communications, " and
"Conducting Business Meetings and Conferences. "

449   Larson, Greta LaFollette.   Business English Essentials.   New
          York: Gregg/McGraw-Hill, 1959.   184p.
This book is divided into four parts.   Parts I-III cover parts of
speech, grammar, and punctuation.   Part IV, "Special Techniques
for Business Letter Writing, " deals with format and types of letters,
such as credit, application, and adjustments.   The parts are further
divided into sections.   Each part has tear-out worksheets for the
student, requiring fill-in responses or asking for writing assignments.

450   Larson, Virginia.   How to Write a Winning Proposal.   Car-
          michael, Calif.:   Creative, 1976.   63p.
This practical guide for writing government and private proposals is
designed for the inexperienced writer.   Set in typewriter font, the
book illustrates and discusses the various parts of proposals.   It
covers in small sections (the longest is five pages) such topics as
job descriptions, proposal outlines, and the proposal review process.
Following the main section are seven appendixes, including a description of a typical private foundation, a sample preliminary proposal form, preapplication form, typical proposal statement, sample
statement of objectives, covering letter, and sample proposal.

451   Lass, A. H.   Business Spelling and Word Power.   New York:
          ITT Educational Services, 1961.   288p.
This workbook with tear-out pages is divided into 24 units.   Each
unit contains a brief explanation followed by two to three pages of
very basic spelling exercises.   The dictionary section, showing words
and definitions, is reprinted from Webster's New World Dictionary.

452   Laster, Ann A. , and Pickett, Nell Ann.   Occupational English.
          2nd ed.   San Francisco: Canfield, 1977.   Index.   328p.
Originally titled Writing for Occupational Education (1974), this text-
workbook is designed for the vocational or two-year student.   To
provide practical help for this student, the 11 chapters begin with
"chapter-opening objectives" and contain plan sheets.   The first five

chapters review basic rhetorical devices (instruction and process explanation, description of a mechanism, definition, analysis through
classification and partition, and analysis through cause and effect).
Chapter 6 covers "the summary." Chapters 7-9 are on forms of
writing--the business letter, the report, and the library paper. Chapter 10 covers oral communication, and Chapter 11, the use of visual
aids.

453 Lawrence, H. C. Turning Him Down: Credit Letters Pertain-
          ing to Declined Orders. 3rd ed. St. Louis: Consoli-
          dated, 1908.

454 Lawrence, Nelda R. Writing Communications in Business and
          Industry. 2nd ed. Englewood Cliffs, N. J.: Prentice-
          Hall, 1974.

455 Lazarus, Sy. Loud and Clear: A Guide to Effective Communi-
          cation. New York: American Management Association,
          1975. 140p.
This book "attempts to analyze some of the commoner breakdowns
[of communication] that affect our everyday interchanges. Its modest
objective is improvement of ability in this universal activity. The
approach is meant to be entertaining." The first three chapters
cover basic communication principles and the ways they are regularly
violated. Chapter 4, "Pardon, My Writing Is Showing," presents
basic principles of good business writing and illustrates the way
these principles are violated, often with humorous results. The final
six chapters discuss nonverbal communication, mass communication,
and the importance of clarity in our everyday and business lives.

456 Lee, James Melvin. Business Writing. Vol. II. New York:
          Ronald, 1920. Index. Bib. 612p.
This is the second volume of a set; the first volume covers business
speech. The 37 chapters in Volume II are divided into seven sections. Part I, "The Essentials of Writing," surveys the basic principles of clear writing. Part II, "The Reinforcement of Reading,"
concentrates on analyzing the writing of others and makes recommendations on business magazines and newspapers. The author suggests
that all readers study Parts I and II, and select from the other parts
according to interest. The other parts review business letters, business reports, advertising copy, and the journalism of business, mechanical and incidental. One chapter covers the house organ. The
bibliography lists handbooks and dictionaries, on both business writing
and business speech.

457 Lee, LaJuana Williams, et al. Business Communication.
          Chicago: Rand McNally College Publishing, 1980. Index.
          Bib. 467p.
The 16 chapters of this textbook are divided into seven parts. Part
I treats the communication process, and Part II, the principles of
effective communication. Parts III-IV are on format, types of letters, research techniques, documentation, and reports. Part V is
concerned with the job search. Part VI covers oral communication

and dictation.   Part VII discusses legal and ethical considerations
and the future of written and oral communication in business.   Part
VII contains ten appendixes, offering such information as business
abbreviations, word processing, addressing envelopes, and numbers
usage.   Chapters contain correction exercises, writing assignments,
and cases.   Chapters also contain photographs and marginal notes.

458   Lefferts, Robert.   Getting a Grant: How to Write Successful
          Grant Proposals.   Englewood Cliffs, N. J. :   Prentice-
          Hall, 1978.   Index.   160p.
The purpose of this book is to guide those seeking grants.   The author says that his principles will cover most grants in human-service
fields, such as education, welfare, employment, and health.   There
are six chapters covering the definition of a proposal, types of proposals, criteria for evaluation proposals, components of a proposal,
and resources for locating funds.   The appendix presents a sample
proposal with a running commentary on format and writing in the
right margin.   Checklists for expenses, capability, and organization
are provided.

459   Leffingwell, W. H.   Automatic Letter Writer and Dictation
          System.   Chicago: Shaw, 1918.

460   Leonard, Donald J.   Shurter's Communication in Business.
          4th ed.   New York: McGraw-Hill, 1979.   Index.   563p.
This textbook is a revision of Written Communication in Business
(3rd edition) by Robert L. Shurter.   The organization and purpose
of this new edition is nearly the same as its predecessor, with two
notable exceptions.   The 22 chapters are divided into four parts, the
fourth covering "nonwritten aspects of written communication" (e. g.,
oral reports and nonverbal communication).   There is also use of
nonsexist pronouns and language throughout the text.   In addition,
the handbook section has been expanded to 114 pages.

461   Lesikar, Raymond V.   Basic Business Communication.   Home-
          wood, Ill. :   Irwin, 1979.   Index.   457p.
This textbook seeks to lighten the teacher's burden with abundant
end-of-chapter exercises, problems, and questions.   The 16 chapters
are divided into six parts.   Part I covers the fundamentals of business writing--adaptation, clarity, and tone.   Part II treats various
patterns in letters: routine inquiries and orders, answering inquiries, and indirectness for bad news and persuasion.   Part III covers
sales, collection, and application letters.   Part IV deals with reports.   Part V explains format, grammar, and punctuation.   Part
VI treats speaking and listening.   The four appendixes include the
answers to a diagnostic test of punctuation and grammar, checklists
for grading letters and reports, and a section on documentation.

462   Lesikar, Raymond V.   Business Communication: Theory and
          Application.   3rd ed.   Homewood, Ill. :   Irwin, 1976.
          Index.   564p.
Designed for advanced students, this book has 16 chapters in two
parts.   Part I, "Communication Theory, " covers such topics as the

role of communication in businesses, a model of the communication
process, and semantics.   Part II, "Applications to Business," reviews the general principles of good business writing, correspon-
dences (indirect and persuasive), reports, style in writing, format,
and oral communication.   Most chapters end with problems or cases.
The five appendixes include format, documentation, grammar/usage
advice, a report checklist, and a general checklist for the instructor
to use in marking letter assignments.

463 Lesikar, Raymond V.   How to Write a Report Your Boss Will
          Read and Remember.   Homewood, Ill.:   Dow-Jones-Irwin,
          1974.   Index.   216p.
This book is directed at young, middle-management executives.   Chapter 1 deals with the purpose of communication, and Chapters 2-3,
with the physical make-up of a report and the patterns of organization.   Chapters 4-6 emphasize writing skills or style appropriate to
technical writing: "Qualities of Effective Writing," "Techniques of
Readable Writing," and "Correctness of Writing."   Chapter 7 covers
graphic aids.   Chapter 8 deals with presentation of reports, and
Chapter 9, with handling data (bibliography).   There are five appendixes giving specimens of a long formal report, a short report, a
memo report, a letter report, and an audit report in a memo form.

464 Lesikar, Raymond V.   Report Writing for Business.   4th ed.
          Homewood, Ill.:   Irwin, 1973.   Index.   389p.
In this fourth edition the author has added a chapter on oral reporting
and more material about research methods.   The book, in 15 chapters, follows the writing process: determining the reader and purpose of a report, collecting information, organizing information,
structuring for various types of reports, writing style, and documentation.   Each chapter ends in discussion questions and writing
exercises and assignments.   Four appendixes are included: a report
checklist, statistical methods, cases, and samples of types of reports.

465 Lesly, Phillip.   How We Discommunicate.   New York: AMACOM,
          1979.

466 Letters That Collect.   Bucyrus, Ohio: Modern Mercantile,
          1910.

467 Level, Dale A., Jr., and Galle, William P., Jr.   Business
          Communications: Theory and Practice.   Dallas: Busi-
          ness Publications, 1980.

468 Levie, Robert C., and Ballard, Lou E.   Writing Effective Re-
          ports on Police Investigations.   Boston: Allyn and Bacon,
          1978.   Index.   424p.
This guide of 13 chapters can be used by "reviewers of police reports, second investigators, and administrators of law enforcement
agencies," as well as attorneys, students and officers.   Part I,
"Writing the Report," includes "Taking Notes During the Investigation," "Analyzing and Organizing the Notes," and "Writing the

Report. "   Part II gives samples of police reports: "Theft Case Report, " "Motor Vehicle Theft Case Report, " "Burglary Case Report, "
with robbery, murder, rape, and traffic-accident reports.   These
samples are thoroughly analyzed.   Chapters contain writing exercises.   The two appendixes provide a checklist for police reports
and a handbook-like treatment of usage, spelling, abbreviations, etc.

469   Levine, Norman.   Technical Writing.   New York: Harper and
        Row, 1978.   Index.   129p.
This text, an expansion of a pamphlet titled "Technical Writing"
(Harper Studies), was designed for introductory college courses in
technical writing.   Part I is the original pamphlet and is a handbook
for grammar, usage, and punctuation.   Part II, "Technical Explanation, " adapts standard expository kinds of writing--definition, description, narration, and inference--to technical writing.   Part III,
"Technical Reporting, " covers the principles, formats, characteristics, and functions of reports.   The 17 chapters use military-numbering
style.

470   Levine, Stanley L.   English: The Power of the First and Last
        Sentences in Business Correspondence.   Tustin, Calif. :
        Media Masters, 1969.

471   Levine, Stuart.   Materials for Technical Writing.   Boston:
        Allyn and Bacon, 1963.   Index.   128p.
This book is a collection of writing assignments called "Jobs. "   They
are designed to "raise the problems which a technical writer might
encounter, and yet which would be completely comprehensible to students in any field (and to the instructor). "   Each of the assignments
has been used in the classroom at least four times, and the book in
mimeograph form was used for three semesters at the University of
Kansas.   The six chapters cover such topics as audience, outline,
clarity, sentence design, graphics, business letters, articles, prospectus, and reports.   Brief explanations are followed by the 27
Jobs, including writing an instruction booklet, writing a letter of advice, and preparing an article for publication.

472   Lewis, Edwin Herbert.   Business English.   Chicago: LaSalle
        Extension University, 1911.   Index.   287p.
Written for students from high school through college, this book covers thoroughly what is now called business writing.   In addition to
chapters on grammar and usage, it includes chapter-length treatments of outlining; the paragraph versus the long sentence; transition (called "connection"); tone, and narration, description, exposition, and argument.   The author also covers business reports, advertising, and letters.   Following is a 76-page section of "Questions,
Themes, and Exercises. "   Nearly every chapter begins with a quotation from or reference to a literary or historical figure and usually
ends with a guiding list of points.   Each new subject introduced
throughout the 24 chapters is numbered--120 such divisions.

473   Lewis, Leslie Llewellyn, ed.   The Business-Letter Deskbook.
        Chicago: Dartnell, 1969.   Index.   Bib.   288p.

This book is composed primarily of samples. Part I, "General Letters, " gives advice on general principles--clarity, persuasiveness,
and word choice, etc. , along with types of letters, stationery and
letterheads, and letters to non-English-speaking countries.   Part II
covers sales letters and salesmanship.   Part III, "Addendum, " is a
collection of general subjects: addressing to an unknown receiver,
openings and closings, typing forms, and proofreader's marks.

474   Lewis, Leslie Llewellyn.   Short Course in Business Corres-
          pondence.   Chicago: Dartnell, 1964.

475   Lewis, Leslie Llewellyn, and French, Marilyn.   New Short
          Course in Business Correspondence.   2nd ed.   Chicago:
          Dartnell, 1965.   219p.
This work, designed as a self-help guide, is three-hole punched for
insertion in a binder; the ten chapters are marked with divider tabs.
The focus is on letter writing, and topics covered include clarity,
conciseness, word choice, letter openings, closings, and shortcuts
to efficiency.   Examples of real letters are included, all on gold-
colored paper for easy reference.   Chapter 10 is "A Self-Check Review, " which contains exercises and answers on vocabulary, word
choice, letter style, usage, and editing.

476   Lewis, Phillip V. , and Baker, William H.   Business Report
          Writing.   Columbus, Ohio: Grid, 1978.   Index.   Bib.
          273p.
This textbook is designed to take students through the nature of reports, research process, and writing process.   The first three parts
cover the report in great depth.   Part IV is on oral reporting.   Part
V presents 14 cases.   Part VI presents 12 articles ranging from how
to develop a presentation objective to communication of financial data.
There is no handbook section.   One section of Chapter 10 explains
that sentences are made up of phrases and clauses.   Readability of
the text is limited to the more sophisticated student because of word
choice and theory description.   A discussion of kinds of "noise, " for
example, explains that "mechanical noise" occurs from spelling errors.

477   Lewis, Richard A.   Annual Reports: Conception and Design.
          Zurich: Graphics, 1971.

478   Lewis, Ronelle B.   Accounting Reports for Management.   En-
          glewood Cliffs, N. J. :   Prentice-Hall, 1957.   187p.
This book "includes business and accounting reports of sales, costs,
expenses, profit or loss, balance sheet elements, cash flow, and
other factors.   It illustrates comparative reports showing current
performance related to the budget and prior year and trend reports
showing past and projected future operations. "   The 13 chapters include the following topics: physical format and "eye appeal, " arrangement of data, quick reports to summarize, monthly accounting
reports, projecting cash requirements, capital expenditures, and
monthly reports to the president.   The book emphasizes ways to
make report writing quicker, easier, and more interesting.

479 Leyton, A. C. The Art of Communication: Communication in
        Industry. London: Pitman, 1968.

480 Lindauer, J. S. Communicating in Business. New York:
        Macmillan, 1974. Index. 502p.
This text, designed for a business communication course, follows a
"single fictitious company in the throes of a contemporary problem:
being caught with an obsolete product and facing change or death."
Otherwise, the content is fairly standard. Part I, "Introduction,"
covers basic communication theory; Part II, "Memoranda"--general
memos, progress reports, and instructions; Part III, "Oral Communication"--interviews and conferences; Part IV, "Reports"--good-news,
persuasive, and bad-news letters. There are writing assignments,
cases, and class-discussion exercises. The two appendixes contain
cases and biographies. A glossary reviews grammar and rhetoric.

481 Lindauer, J. S. Writing in Business. New York: Macmillan,
        1971. Index. 269p.
This book stresses communication as the exchange of meaning between or among human beings "by way of one of numerous channels."
It emphasizes the necessity and value of self-knowledge, control over
subject matter, and analysis of the reader. The 17 short chapters
on letter writing are followed by a 79-page glossary of grammatical,
usage, and rhetorical practices. Chapter titles include "Purposes
of Letters," "Maintaining an Image," "Conventions of Content,"
"Good-News Messages," "Persuasive Messages," and "Letters as
Reports." Chapter 17, "A Checklist," is a two-page summary of
the principles in the book. Each chapter is followed by discussion
questions and writing problems.

482 Linton, Calvin D. Effective Revenue Writing 2. rev. ed.
        Training 9931-15. Washington, D.C.: U.S. Government
        Printing Office, 1978. Index. 198p.
This book is a guide to writing based on the author's lectures in an
advanced course for government writers. The basic course, Effective Writing I, covers grammar fundamentals. This course stresses
structure. The author states that "the necessity of practice may be
met in several ways, but perhaps the most practical is to consider
each of your normal writing jobs as an opportunity for practice."
The contents are aimed directly at, "but ... not exclusively," employees of the Internal Revenue Service. The author also refers to
the book as a "course for the correspondence student." There are,
however, no exercises or self-study devices. Eleven chapters,
"written lectures," are on such topics as "Nature and Function of
Written Communication," "Diagnosing Sentence Difficulties," "The
Syntax of Strong Sentences," "Economy in Writing," and "Style in
Expository Writing." A glossary contains examples of grammar
terms and specialized terms used in the lectures.

483 Linton, Calvin D. How to Write Reports. New York: Harper
        and Brothers, 1954. Index. 240p.
This book of 13 chapters is designed for people on the job. Part I,
"The Report," reviews general communication theory and the need

for clear writing.   Part II, "The Methods," covers research, sentences, paragraphs, and style.   Part III, "The Mechanics," is a
brief, selective overview of grammar and punctuation.   A chapter
on "Visual Representation" focuses on graphics in reports.   The
glossary covers grammar and stylistic terms.

484 Linton, Marigold.   A Simplified Style Manual: For the Prep-
         aration of Journal Articles in Psychology, Social Sciences,
         Education, and Literature.   New York: Appleton-Century-
         Crofts, 1972.   Index.   Bib.   184p.
Despite the scope suggested in the title, this manual is "designed
primarily for people preparing ... to publish in American Psychological Association (APA) journals .... "   The book, then, is a simplified version of the APA style guide.   The eight chapters cover style
and the structure of a research paper (title, abstract, introduction,
method section, results section, discussion, and references).   The
final chapter provides a sample article--first in typewritten form
and then in printed form as it appeared in a journal.   The appendix
consists of five parts: Metric System, Abbreviations, Spelling, References on Style, and Proofreader's marks.

485 Little, John D.   Complete Credit and Collection Letter Book.
         2nd ed.   Englewood Cliffs, N.J.: Prentice-Hall, 1964.

486 Lloyd-Jones, Richard, and Andrews, Clarence A.   Technical
         and Scientific Writing.   Iowa City: Department of Eng-
         lish, State University of Iowa, 1963.   183p.
This spiral-bound textbook is written for those students who wish to
have a career in technical or scientific writing; the authors stress
also that the subject is "the classical skills of rhetoric and exposition" applied to modern technical and scientific subject matter.
Section I, "Method," discusses the application of classical rhetorical
order to technical subject matter.   Section II, "Exposition," examines
classical rhetorical forms as they apply to technical subjects--including
description of a mechanism.   Section III, "Persuasion," covers various methods of rhetorical argument.   Section IV is "Style."   Section
V, "Visuals," briefly reviews types of illustration.   The appendix
gives some samples of classical technical writing and forms of reports.

487 Lock, Stephen Penford.   Thorne's Better Medical Writing.   2nd
         ed.   Kent, England: Pitman Medical, 1977.   Index.
         Bib.   118p.
This book is written for the physician intending to publish.   The author, editor of the British Medical Journal, wrote the first edition
with Tony Smith under the name of an Anthony Trollope character,
Dr. Thorne.   The 18 chapters take the reader from "When to Start"
to the steps of proofreading and editing.   Three chapters are on
"Better Scientific Style."   Although the book is British, one chapter
discusses "Choosing Among Some American Journals," and another
chapter covers "How Can Foreign Authors Help Themselves?"   The
five appendixes include further readings, a discussion of medical
ethics, and words to avoid.

488   Locke, Lawrence F., and Spirduso, W. W. Proposals That
         Work. New York: Teachers College, 1976.

489   Lockley, Lawrence C. Principles of Effective Letter Writing.
         2nd ed. New York: McGraw-Hill, 1933. Index. Bib.
         440p.
The 24 chapters in this textbook are divided into three parts. Part
I covers "Fundamental Principles" of letter writing: ease of reading, tone, and dictation. Part II deals with different kinds of business correspondence: letters of inquiry, adjustment, credit, collection, etc. Part III deals exclusively with "Selling by Mail," showing
ways of transmitting promotional material. The appendixes cover
"Letters and the Law," "Writing Reports," "The Form of the Letter,"
"Reducing the Cost of Letters," "Exercises in Correct English," and
a bibliography. Each chapter ends with problems for discussion and
writing/revising assignments. There are many sample letters to be
used as models.

490   Long, Sandra Salser. Transmission: Communication Skills for
         Technicians. Reston, Va.: Reston, 1980. Index. 388p.
This text-workbook, in 16 chapters divided into four parts, is designed
specifically for vocational students. As suggested by its title, this
book uses automobile and machine terms to explain communication
principles. Part I, "The Communication Machine," discusses methods and problems of communication. Part II, "Operational Analysis,"
covers the process of communication related to written and oral
communication. Part III, "Functions of the Mechanism," treats the
purposes of communication: explaining, reporting, persuasion, and
problem solving. Part IV, "Troubleshooting Language," provides a
series of competency tests and practice with parts of speech, punctuation, and spelling. Part V, "Competency Based Duties," provides
nine types of communication assignments, such as filling out job applications and other printed forms, writing letters, preparing visuals,
and researching and writing a formal report. Students complete all
tests and exercises in the workbook, which has tear-out pages.

491   Loomis, H. T. New Practical Letter Writing. Cleveland:
         Practical Textbook, 1911.

492   Lytel, Allan. Technical Writing as a Profession. Cincinnati:
         privately printed, 1959. Index. Bib. 113p.
This book aims to "aid and encourage the growing profession of technical writing." The author first defines the field of technical writing,
finding its roots in early documents but emphasizing its professional
genesis in World War II. The 11 chapters cover such topics as the
nature of the technical writing job, education of technical writers,
and forms of technical writing (engineering reports, instruction manuals, advertising, scientific journals, technical journalism, and
books). Also covered is the technical writing team: illustrators,
editors, printers, writers, and librarians. An appendix contains
lists of professional societies, journals, college courses, books,
and technical-book publishers.

493 Lytle, John Horace. Letters That Land Orders or How to
       Make Letters Sell Goods. New York: Ronald, 1911.
       170p.
The author wrote this book in the "firm belief that it will be of in-
estimable value." The first three chapters stress principles of effective sales letters based on numerous sample letters in which the
author has inserted fictitious names. Chapter 4, "Enclosures and
Postage," discusses at length the procedure for the stenographer
typing abbreviations of enclosure items and the procedure for the
clerk who stuffs the envelope. Other advice is that "it is quite
universally granted that it takes a two-cent stamp to reach the busy
business man. It is likewise quite sure that a one-cent stamp will
reach the farmer who doesn't get much mail." Chapter 5 covers
"The Follow-Up File," and Chapter 6 deals with "Paragraphs and
Punctuation." The author advises writers to collect a set of "best
paragraphs," index them, and use them in future letters. As the
system becomes perfected, "dictating can be eliminated" since the
secretary will know the paragraphs by heart. Chapter 7 deals with
collection letters. The last chapter, Chapter 8, presents "Miscellaneous Sample Letters."

494 McCartney, Eugene S. Recurrent Maladies in Scholarly Writ-
       ing. Ann Arbor: University of Michigan Press, 1953.
       Index. Bib. 141p.
The author, retired editor at the University of Michigan Press,
makes an articulate case for logic, simplicity, and clarity in scholarly writing. This is not in a strict sense a "how-to" book; rather,
it presents various problems in scholarly writing and allows readers
to compare their writing with the standard the author presents. Topics include simplicity of sentences and word choice, logic in the organization of papers, wordiness and tautology, and consistency of
data. The book is liberally illustrated with passages from manuscripts submitted to the author and cartoons from periodicals, such
as The New Yorker and Colliers.

495 MacClintock, Porter Lander. The Essentials of Business Eng-
       lish. Chicago: LaSalle Extension University, 1915.
       Index. 273p.
The author states that this book requires that students have had six
or seven years of elementary education. The 12 chapters cover
grammatical exactness, correct diction, and punctuation. Chapter
9 deals with speech. Chapter 12, "Business Composition," is divided
into sections on business letters, miscellaneous business documents,
and social letters. The numerous exercises cover grammar and
punctuation or require the student to write letters dealing with a
specified problem.

496 McCloskey, John C. Handbook of Business Correspondence.
       1932; rpt. New York: Prentice-Hall, 1941. Index. 467p.
This reference book has 14 sections covering such topics as punctuation and capitals, words, grammar, sales letters, letters to special
classes, credit letters, and application letters. Each section has

eight to 13 subsections covering separate aspects of the topic.   The
section "Letters to Special Classes" has a subsection on "letters to
women" and one on "letters to men. "   The advice to someone writing
to a man is to "base letters to men on logic and conviction"; a writer
to a woman should "use persuasion and emotional appeals rather than
logical argument. "

497   MacDonald, G. R.   Pitman's Manual of Spanish Commercial
          Correspondence.   New York: Pitman, n. d.

498   McIntosh, Donal W.   Techniques of Business Communication.
          2nd ed.   Boston: Holbrook, 1977.   Index.   430p.
This textbook of 15 chapters covers business communication, focusing
on the ultimate psychological purpose of business writing.   For example, Chapter 2, "Developing Empathy, " and Chapters 9 and 11,
"Writing with Feeling" and "Selling Yourself, " all focus on the psychology of business writing as well as the function.   Other chapters
present treatments of reports, letters, and editing.   There are writing assignments, cases, and discussion questions for the student.

499   McJohnston, H.   Business Correspondence.   New York: Alex-
          ander Hamilton Institute, 1917.

500   McLaughlin, Ted J. ; Blum, Lawrence P. ; and Robinson, David
          M.   Cases and Projects in Communication.   Columbus,
          Ohio: Merrill, 1965.   117p.
According to the authors, this workbook with tear-out pages "begins
where our book Communication concludes. "   It provides a wide variety of case studies, with blanks for the student to fill in responses.
It is "self-contained" in that it can be used as an independent workbook because "it contains exposition of communication principles and
practices; in addition, each case or project is introduced by a statement of purpose and accompanied by specific instructions. "   Most
of these cases are in a business context.   Part I, "Foundations of
Communication, " is a brief treatment of general psychological principles and the application of those principles to business situations
and management decisions.   Part II, "Person-to-Person Communication, " covers one-to-one oral communication and individual communication through letters (informative, inquiry/request, persuasive request, goodwill/sales, "bad news, " and collection).   Part III, "Group
Communication, " covers group speaking and business reports.

501   McLaughlin, Ted J. ; Blum, Lawrence P. ; and Robinson, David
          M.   Communication.   Columbus, Ohio: Merrill, 1964.
          Index.   499p.
This book of 20 chapters emphasizes "the unity of communicative
elements within a diversity of forms. "   Part I, "Foundations of
Communications, " covers human psychology and the communication
process.   Part II, "Person-to-Person Communication, " discusses
memos and business letters (routine and persuasive) in addition to
interviews, conferences, personnel problems, and maintaining discipline within an organization.   Part III, "Group Communication, "
covers reports, training programs, and management publications as

well as improving communication within the management group. Part IV, "Communication with Outside Groups," covers communication for management to various public and labor unions. The chapters end with summaries, "projects for study and practice" (discussion, re- search and writing exercises), and suggested readings.

502 Maclean, Joan. English in Basic Medical Science. London: Oxford University Press, 1975. 112p.
This textbook is for students who are non-native speakers of English. The eight units are on such topics as the heart, cell structure, and the nervous system. The units consist of brief readings and com- prehension questions followed by exercises in sentence completion and revision and short paragraph assignments. Answers to the com- prehension questions are included. Units 1-7 end in a free reading passage. Unit 8 ends in a lengthy list of notes. The student is to write an essay from the notes.

503 McNaughton, F. More Business Through Postcards; An Ex- haustive Analysis of Possibilities for Intensively Increas- ing Profitable Sales Through Return Postcards. Chicago: Selling Aid Cut Service, 1917.

504 MacRae, James. How to Write Love Letters to Policyholders. Chicago: Aldine, 1965.

505 Mager, N. H. , and Mager, S. K. The Complete Letter Writer. rev. ed. New York: Pocket Books, 1968. 310p.
This book of model letters is divided into two parts: "Social Letters" and "Business Letters." The 11 chapters that deal with business letters cover such topics as "Sales Letters," "Credit Letters," "Routine Business Letters," and "Turning Inquiries into Sales." Each of the 27 chapters consists of tips for a specific letter type and then samples of letters. Some chapters contain lists of helpful starting sentences or key sentences for the topic. The chapter on "Good-Will Letters" has a list of starting phrases.

506 Maizell, Robert E. ; Smith, Julian F. ; and Singer, T. E. R. Abstracting Scientific and Technical Literature, An Intro- ductory Guide and Text for Scientists, Abstractors, and Management. New York: Wiley-Interscience, 1971. In- dex. Bib. 297p.
The chapters in this guide define and explain types of abstracts and abstracting services. Included are instructions on writing abstracts as well as operating an abstracting program. One chapter explains the function of the computer. There are ample illustrations.

507 Makay, John J. , and Fetzer, Ronald C. Business Communica- tion Skills: A Career Focus. New York: Van Nostrand, 1980.

508 Mandel, Siegfried. Writing for Science and Technology: A Practical Guide. New York: Dell, 1970. Index. 353p.
This book begins with an overview of the profession of technical writing. Chapters 2-5 cover usage in technical writing, writing

definitions, editing techniques, and outlining. Chapters 6-9 deal with
technical reporting and reports--the writing process, abstracts and
summaries, and formats and conventions. Chapters 10-11 review
short forms: the news release, memo, and letter. Chapters 12-13
cover instruction, specifications, and proposals. Chapter 14 is a
style guide. The appendix contains four articles about writing. Nearly two-thirds of the book is devoted to examples.

509   Mandel, Siegfried.   Writing in Industry.   Vol. I.   Brooklyn,
            N. Y. :   Polytechnic, 1959.   121p.
This book is a collection of seven presentations given at a conference
held at the Polytechnic Institute of Brooklyn.   The presentations are
"The Challenge to Writers in Industry," by Siegfried Mandel; "The
Relationship of Engineering and Technical Writing," by Robert T.
Hamlett; "Everyday Editorial Problems of an Engineer-Supervisor,"
by Ronald J. Ross; "Techniques and Practices of Proposal Writing,"
by David L. Caldwell; "Writing for Publication: Why and How?,"
by George R. Wheatley; "Production and Design Problems in Engineering Publications," by Arthur Eckstein; and "Journalistic Aspects
of Science Writing," by William L. Laurence.

510   Mandel, Siegfried, and Caldwell, David L.   Proposal and In-
            quiry Writing.   New York: Macmillan, 1962.   Index.
            246p.
The eight chapters of this book are arranged roughly in the sequence
of the proposal process, covering the initial inquiry; structure, content, and language of the proposal; printing and physical production
processes; and outlining, writing style, and mechanics.   Chapter 8,
"The Contract," covers preparing the contract after the bid has been
accepted.   Chapters 3-4 include a sample proposal with analysis,
an outline of the proposal flow, and a description of a typical proposal team.   Four supplements illustrate a letter of transmittal and
introduction; a checklist of the steps of the proposal process; an
abridged version of a proposal; and two techniques used to plan and
evaluate the development of a complex weapon system: Program
Evaluation Review Technique (PERT), used by the Navy, and Program Evaluation Procedure (PEP), used by the Air Force.

511   Manly, John Matthews, and Powell, John A.   Better Business
            English.   Chicago: Drake, 1921.   Index.   217p.
Coauthored by the Head of the Department of English at the University of Chicago, John Manly, this book is designed for "writers who
wish to find immediate replies to the many puzzling questions about
spelling, the use of capitals, ... the meanings and uses of words,
grammatical correctness, and the construction of sentences and paragraphs." However, the first two chapters deal rather philosophically
with the art and process of writing and "what is good English?"
Chapter 9 is on the format of business letters.

512   Manly, John Matthews, and Powell, John A.   Better Business
            Letters.   Chicago: Drake, 1921.   Index.   Bib.   167p.
This book is designed "as a practical aid to the man who is trying
to train himself to break away from the stiff commonplace office

letter that is 'natural' to him only because it has become habitual. "
The eight chapters cover such topics as the format of sales letters,
answering complaint letters, the importance of the comparative study
of letters, follow-up letters, collection letters, and the psychological
value of style.

513 Mann, Charles. Editing for Industry: The Production of House
        Journals. London: Heinemann, 1974. Index. 188p.
This book is a working reference for an industrial editor and a sup-
plement manual for the student taking a course through the British
Association of Industrial Editors. The 16 chapters cover such topics
as "What Kind of House Journal?, " "Content, " "Clarity Through
Grammatical Accuracy, " "The Editor and Management, " "Sub-Editing, "
"Typography, " and "Complying with the Law. " The glossary covers
printing and technical terms.

514 Marcoux, Harvey Lee. A College Guide to Business English.
        New York: Van Nostrand, 1939. Index. 591p.
The premise of the textbook is that "the student cannot learn to write
business English until he has learnt to write English. " The first of
five sections, therefore, reviews sentence structure and grammar.
Section II covers types of business letters (credit, sales, adjustment,
etc.). Section III deals with the mechanics of gathering data, writing
the draft, and the business report. Section IV is devoted to "Advertising Copy. " Section V deals with oral reporting. The five appendixes include topics for reports and outlines of types of business reports.

515 Marder, Daniel. The Craft of Technical Writing. 2nd ed.
        Dubuque, Iowa: Kendall/Hunt, 1976. 225p.
The author laments in his preface that few books published since the
first edition of this text (1960) have attended to the real business of
writing. Most of the five sections in this edition cover exposition
and style. Section I, "The Craft, " presents a comparison between
scientific exposition and basic rhetorical approaches. Section II,
"Techniques of Organization, " covers paragraphs and methods of development. Section III, "Structure of the Whole, " deals with techniques for beginning, developing, and ending. Section IV, "Techniques of Style, " covers sentences, language, and the mechanics of
style. Section V, "Some End Products, " shows examples of various
forms: formal report, proposal, memo, and article.

516 Marti-Ibanez, Felix, ed. Medical Writing. New York: MD
        Publications, 1956.

517 Martin, George W. Let's Communicate: A Self-Help Program
        on Writing Memos and Letters. Reading, Mass.: Addison-
        Wesley, 1970. Bib. 125p.
This textbook, with tear-out pages, is designed to be used in the
classroom or individually by students or business people. Each of
the seven units discusses a principle, then follows with "Spot Check:
Questions, " "Spot Check: Answers, " "Rewrite Assignments, " and
"Possible Rewrite Assignment. " Topics covered are the importance

of good writing habits, planning, conciseness, simplicity, writing
style, openings and closings, sentences and paragraphs, and positive
tone. Unit VII gives a review and advice on editing. There are
three appendixes: a list of selected books for study, a "Let's Communicate Pledge, " and a 17-point "Checklist for Evaluating Memos
and Letters. "

518 Martin, Roy. Writing and Defending a Thesis or Dissertation
in Psychology and Education. N. p. , 1980.

519 Marting, Elizabeth; Finley, Robert E. ; and Ward, Ann, eds.
Effective Communication on the Job. rev. ed. New
York: American Management Association, 1963. Index.
304p.
Drawn from both business and academe, this collection of essays
on business communication aims to give business professionals practical, immediate help in solving communication problems, from reprimands to employees to giving the boss what he or she needs to
know. There are five sections. Section I, "Bridges and Barriers, "
contains two articles on the problem of semantic ambiguity in managing. Section II gives advice on self-improvement, and Section III
reviews upward, downward, and lateral communication. Section IV
covers giving instructions, evaluating employees, and improving
meetings. Section V provides articles about writing letters and
reports--by recognized names in business communication.

520 Masterman, James R. , and Phillips, Wendell Brooks. Federal
Prose: How to Write in and/or for Washington. Chapel
Hill: University of North Carolina Press, 1948. 45p.
This book is a humorous approach to gobbledygook in government
writing. The ironic slant is that the prose supposedly being
praised while it is really being ridiculed. Numerous cartoons add
to the message. The book is organized around six questions, each
covered in a chapter. Questions include "Have You the Courage to
Write Federal Prose?" and "Have You Any Aptitude for Writing
Federal Prose?"

521 Masterman, L. E. The Mechanics of Writing Successful Federal Grant Applications. Columbia: University of Missouri Press, 1973.

522 Mathes, J. C. , and Stevenson, Dwight W. Designing Technical
Reports: Writing for Audiences in Organizations. Indianapolis: Bobbs-Merrill, 1976. Index. 396p.
This textbook is aimed at engineering students or young professionals,
giving them "a systematic procedure which will enable the engineer
to approach and solve the problem of report design confidently and
efficiently. " The ten chapters are divided into three parts: "Determining the Function of the Report in the System, " "Designing the
Report, " "Writing and Editing the Report. " Topics include audience
analysis, report structure, organization, editing, and layout. There
are numerous illustrations and an 18-page section of guides and
checklists keyed to the chapters. The nine appendixes contain "nine

umé2

complete reports--all written on the job by professionals," ranging from one page to 50 pages, including a résumé and job application.

523 Matthews, Lempi K. Making the Most of Your Annual Report. Chicago: Public Personnel Association, 1963. Bib. 76p. This work contains suggestions for making annual reports appealing to readers. The topics include "What Should Go into the Annual Report," "How to Say It: Writing Style," "Handling Layout and Content," "Other Ways to Tell Your Story," and "Putting the Annual Report to Work for You." The "Special Supplement" covers tips for layout, artwork, and marking copy.

524 Matthies, Leslie H. Playscript Procedure: A New Tool of Administration. New York: Office Publications, 1961. Index. 183p. This book gives advice for writing clear administrative procedures. The author stresses the importance of readable procedures. The book is divided into eight "Acts," which include "What Is a Good Procedure?"; "Research in Procedures Communication"; "People, Playscript, and Action"; and "Common Errors in Procedures." The book suggests the playscript format, actors (employees) in a drama (business situations). The procedure is shown in two columns: Responsibility (who does it) and Action (what is done). Cartoons are used throughout.

525 Maude, Barry. Practical Communication for Managers. London: Longman, 1974. Index. 217p. This book deals with the manager's job in all areas of communication and the barriers to that communication. The 16 chapters include checklists and self-quizzes to help students or managers absorb the advice in the chapter. Topics include managing by meetings, influencing committees, and public-speaking media. Chapters 6-8 deal directly with writing memos, letters, reports, and company publications. There are three appendixes. Appendix I is a two-page discussion of conference attendance. Appendix II, "A Note on Grammar and Punctuation," is a one-page guide in which the author advises the use of the dash "to save the writer the trouble of choosing the precise punctuation mark that is needed." The author comments that "grammar is useful only if ... it enables accurate communications to be passed." Appendix III is a short discussion of the value of listening.

526 Mavor, W. Ferrier. English for Business. New York: Beekman, 1974.

527 Maybury, Sally B. Principles of Business Letter Writing. New York: Ronald, 1959. Index. 413p. This ten-chapter textbook was written "primarily for use in business letter writing courses in colleges, business schools, and secretarial schools." Each chapter contains three or more pages of various exercises, cases, and writing assignments. The book is divided into four sections. Part I, "Fundamental Principles," contains two

chapters on business letters.   Part II, "Writing the Message, " contains two chapters on style.   Part III, "Presenting the Message"
contains two chapters on letter form.   Part IV, "Using the Principles, " contains four chapters on types of letters: daily and "im-
plementing" letters; sales letters; collection and claim letters, and
application letters, including résumés.   The appendix contains six
reference sections: a short usage glossary, a list of dictionaries,
a section on spelling/punctuation, a grammar section, a brief description of types of paper, and a list of special forms of address
and salutation.

528   Mayer, Edward N. , Jr.   How to Make More Money with Your
          Direct Mail.   3rd ed.   Pleasantville, N. Y.:   Printers'
          Ink, 1957.   Index.   363p.
This book is intended as a guide to the beginner and a reference tool
for the expert.   There are 32 chapters divided among six parts.
Part I, "What You Should Know at the Start, " defines and explains
the function of direct mail, the types of direct mail, and lists rules
for success.   Part II, "What You Put into Direct Mail, " discusses
planning, good letters, attention getting, and the personal touch.
Part III, "How Your Direct Mail Looks, " covers appearance, printing, and stationery, along with explaining the 21 types of direct-
mail letters.   Part IV, "Where Your Direct Mail Goes, and How, "
explains the use of a mailing list.   Part V, "How You Can Get Better Results, " discusses achieving interest and continuity in a campaign and techniques in addressing.   Part VI is on "How to Save
Money. "

529   Mehaffy, Robert E.   Writing for the Real World.   Glenview,
          Ill. :   Scott, Foresman, 1980.   Index.   378p.
This text-workbook is designed for students "who need an introductory
course prior to taking university level technical writing courses. "
The 16 chapters are divided into five parts.   Part I covers sentence
construction; Part II, short forms of on-the-job writing--memos,
letters, résumés, and standard forms; Part III, paragraph development and organization; Part IV, reports and college papers; and
Part V, punctuation, spelling, word choice, capitalization, and basic
sentence errors.   Each chapter contains fill-in exercises and writing
problems.

530   Melrose, John.   Bucomco: A Business Communication Simula-
          tion.   Chicago: Science Research Associates, 1977.
          172p.
This textbook posits a fictitious company in which the students become the employees.   The entire textbook is devoted to establishing
this company and presenting simulated situations to which the students respond.   The teacher acts as the president, who chooses
various situations from a "simulator" section to challenge the students.   The book includes job descriptions for the student/manager
positions, e.g., vice president, marketing manager, and credit manager.   An imaginary executive director sends memos to which the
students must respond.   The students, therefore, are "actively involved in speaking, writing, problem solving, and decision making. "

The simulation is designed to fill an entire semester, but the author suggests that it can be used with a standard text as well.

531 Menning, J. H.; Wilkinson, C. W.; and Clarke, Peter B.
      Communicating Through Letters and Reports.   6th ed.
      Homewood, Ill.: Irwin, 1976. Index. 655p.
Although this textbook has been reduced in size in the 6th edition, it covers topics from the psychology of business letters to the value of using colored stationery. It also includes numerous examples, cases, and illustrations. The text is divided into three parts. Part I discusses style, the psychology of business letters, and appearance. Part II covers job procurement and types of letters, including neutral and "good news," disappointing, and persuasive messages. Part III examines several types of long and short reports. The three appendixes cover communication theory, semantics, and dictation and include a concise handbook treatment of grammar and mechanics. Also included are detailed checklists for practically every letter and report situation discussed.

532 Menzel, Donald H.; Jones, Howard Mumford; and Boyd, Lyle
      G. Writing a Technical Paper. New York: McGraw-
      Hill, 1961. Index. Bib. 132p.
This book of seven chapters "attempts to give practical help ... to all ... who want to write as well as possible about some aspect of science or engineering." Chapter 1, "The Evolution of a Paper," gives a view of methods of composition processes to produce a first draft. Chapter 2, "Revision," examines audience, structure, phrasing, and documentation. Chapter 3, "Presenting the Data," covers using symbols, equations, and math. Chapters 4-6 deal with grammar, usage, style, and jargon. Chapter 7, "The Physical Manuscript," covers the mechanics of submission: style sheets, copy preparation, and illustration. Following the chapters are a brief bibliography and an appendix with specific advice on the preparation of the thesis, monograph, and contract report.

533 Meredith, Patrick. Instruments of Communication: An Essay
      on Scientific Writing. London: Pergamon, 1966. Index.
      Bib. 645p.
This ambitious work by a professor of psychology at the University of Leeds examines the process of scientific writing from the perspective of epistemology, psychology, linguistics, semantics, psychophysics, and pragmatics. This study analyzes the scientific mind and method as they shape the writing of scientific and technical information. Part I, "On the Communication of Understanding," analyzes the epistemology of prose, science, and logic. Part II, "On the Meanings of Science," examines the influence of semantics on communication, comprehension, and philosophy in science. Part III, "On Forms of Representation," assesses syntactics in writing, graphics, mathematics, and thought. Part IV, "On the Instrumentality of Language," examines the relationships of psychophysics to scientific and technical writing. In Chapter 13, "The Grammar of Science," the author points to the practical value of the theoretical study and teaching of scientific writing. Part V, "On the Writing of Science,"

is directly concerned with the results of his investigation, on language and writing especially. He points out that when scientific reports do not admit their subjectivity, the "writer clouds his own communication by presenting phenomena full of the shadows of ignored variations. The comprehensibility of his report inevitably suffers. The reader obscurely knows that experiments do not just happen. They are designed." This statement represents one conclusion of the cumulative effects of many factors on scientific writing. The work concludes with an appendix: "On the Syntax of Concepts"; an "Index of Authors and References"; "Publications on Science, Education, Psychology and Communication"; "Literary Acknowledgements"; and a "Topic Index."

534 Mergel, Margaret Z. Communication for Business. N. p. , n. d.

535 Methold, I. K. , and Waters, D. D. Understanding Technical English. Hong Kong: Longman, 1973.

536 Methold, Kenneth. Practice in Medical English. New York: Longman, 1975.

537 Miles, Dudley. English in Business. New York: Ronald, 1920. Index. 449p.
"This book is for boys and girls who are going out into active life rather than for those who go on to college. It will meet the needs of those commercial high schools that train exclusively for a business career." It is also meant for high schools with "commercial departments" and for "all those continuation and corporation schools that give training to boys and girls who have gone to work before securing a high school diploma." The 34 chapters are divided into three sections. Section I, "Expressing Ideas in Elementary Business Situations," contains chapters on speech, grouping thoughts, writing unified and coherent sentences, diction, and writing remittances. Section II, "Expressing Ideas for Effective Business Communication," is devoted to the letter (claim and adjustment, credit, sales, advertising). Section III, "Expressing Ideas Clearly and Correctly," contains chapters on grammar and usage. There are six appendixes: a glossary of usage, a listing of states and territories in the U. S., a glossary of abbreviations, and directions for sending telegrams, for filing, and for proofreading. Each chapter has exercises with writing assignments or essay questions.

538 Miller, Walter J. , and Saidla, Leo E. A. , eds. Engineers as Writers: Growth of a Literature. Essay Index Reprint Series. Freeport, N. Y. : Books for Libraries, 1953. Index. Bib. 340p.
This book is a collection of essays by well-known engineers of the past--from Frontinus' classic report on the aqueducts of Rome to Herbert Hoover's "Report on the Mississippi Flood." According to the preface, "with this book, students can learn about the types, forms, problems, and standards of engineering writing by reading and discussing some of the professional literature itself." Each article contains an introduction by the editors, a comment at the end, and suggestions for further study. A subject index is included.

539   Mills, Gordon H., and Walter, John A.   <u>Technical Writing.</u>
            4th ed.   Holt, Rinehart and Winston, 1978.   Index.   Bib.
            587p.
This textbook is aimed at four-year college students of technical
writing as well as practicing scientists and engineers.   Section I
covers outlines, abstracts, and technical writing.   Section II reviews
rhetorical strategies, such as definition, process, classification, and
interpretation.   Section III deals with introductions, conclusions, and
transitions.   Section IV discusses types of reports: oral and written.
Section V deals with report layout, and Section VI, with the library
research report.   Each chapter has detailed suggestions for writing
assignments.   There are seven appendixes: (A) a selected bibliography, (B) grammar and usage, (C) excerpts from a report to illustrate a typical technical writer's practical problems of communicating
with the reader, (D) organizing the research report, (E) two versions
of a student report, (F) metric-conversion tables, and (G) approved
abbreviations of scientific and engineering terms.

540   Mills, John.   <u>The Engineer in Society.</u>   New York: Van Nos-
            trand, 1946.   196p.
The author believes that engineers "need to learn to write so that an
ordinary person could understand."   The first four sections of the
book discuss personal qualities of scientists and engineers, their
roles as managers in business, and their need to set professional
directions for the future.   The final section, "Exposition for Engineers," is composed of six chapters that discuss the process of
composition.   Chapter 17, "Euclid and King James," relates logical
thinking to clear writing.

541   Milton, Hilary H.   <u>Steps to Better Writing.</u>   College Park,
            Md.:   Professional Press, 1959.   98p.
This book presents topics related to writing for business, government, and industry.   Part A, an overview of the writing process,
is 23 pages long.   Part B contains the five chapters of the book,
covering such topics as organizing and outlining, transitions, introductions, word choice, sentence construction, illustrations, and editing.   Included is a "Checklist for Writers."

542   Mitchell, John.   <u>A First Course in Technical Writing.</u>   London:
            Chapman and Hall, 1967.   Index.   Bib.   180p.
This British textbook is interesting because it "covers the whole of
the syllabus of the City and Guilds Technical Writing Certificate (329)
examination."   The 26 chapters are divided into seven parts.   Part
I begins with an overview of technical writing, its importance and
content.   Part II covers collecting information, and Part III covers
the writing process and style as well as reader adaptation.   Part IV
covers illustrations; Part V discusses the technical report.   Parts
VI and VII cover oral presentation and such forms as letters, memos,
instruction, and sales leaflets.   There is a glossary of English usage, exercises for class, and writing assignments.

543   Mitchell, John.   <u>Handbook of Technical Communication.</u>   Bel-
            mont, Calif.:   Wadsworth, 1962.   Index.   Bib.   321p.
This textbook is for "courses in technical writing and advanced ex-

position" for science and engineering students.   The book has 25
chapters in four parts, covering such topics as instructions, process
descriptions, formal technical reports, proposals, business letters,
speeches, and articles.   Also covered are copy preparation (including
reference sections on style, mechanics, and printing) and legal and
moral aspects of writing (including contracts, patents, and copyright).

544   Mitchell, John.   How to Write Reports.   Glasgow, Scotland:
            Fontana, 1974.   Index.   157p.
This British book focuses primarily on technical-report writing.   Section I, "Introduction," contains a one-sentence definition of the technical report and a two-page discussion of the need for reports.   Section II, "Producing a Report," covers reader analysis, layout, writing, revision, editing, working with the typist or printer, illustrations, etc.   Section III, "Common Kinds of Reports," covers such
topics as form, lab, and management reports, and minutes.   This
section also gives advice on writing reports about people, with a
vocabulary list of positive and negative attributes (e.g., "active--
apathetic," and "dashing--ordinary").   Section IV, "English for Report Writers," discusses effective language, conciseness, clarity,
sentences, paragraphs, and punctuation.

545   Mitchell, John.   Writing for Professional and Technical Jour-
            nals.   New York: Wiley, 1968.   405p.
This book serves the beginning technical writer as well as the professional.   The first four chapters include a thorough discussion of
preparing to write a technical paper; collection, correlation, and selection of data; and organization.   Chapter 5 is an anthology of rep-
resentative styles and articles that is arranged by discipline or subject.   An appendix of abbreviations is included.

546   Monaghan, Patrick C.   Writing Letters That Sell: You, Your
            Ideas, Products and Services.   New York: Fairchild,
            1968.   186p.
This book of 17 chapters covers three major types of letters: personal-
business, consumer, and commercial.   Chapter topics include, for
example, resignation letters, personalized fundraising letters, letters
selling the speciality, closings, letters for smaller retailers and
service stores, letters to employees, letters to stockholders, and
how to decide which officer to address.

547   Monro, Kate M., and Wittenberg, Mary Alice.   Modern Busi-
            ness English: A Text-Workbook for Colleges.   5th ed.
            New York: Gregg/McGraw-Hill, 1972.   Index.   218p.
This spiral-bound workbook is a comprehensive review of grammar
and punctuation divided into nine parts.   The first half of the book
is divided into units covering such topics as verbs, pronouns, capitalization, and numbers.   Unit XXXIV is "The Business Letter."   The
second half of the book consists of exercises on tear-out worksheets
geared to the units.

548   Monroe, Judson.   Effective Research and Report Writing in
            Government.   New York: McGraw-Hill, 1980.   Index.

Bib.   289p.

According to the author, there are "two certainties in government work--change and reports. " This book, designed to help those in government research and write reports, is composed of 18 chapters divided into six sections. Section I covers the politics of government reports and organization; Sections II-III, audience analysis, brainstorming, outlining, research, and collecting data; Sections IV-V, the interpretation of data and writing and editing the report. Section VI explains the management of research. Chapters are illustrated with writing samples and end with summaries.

549   Monroe, Judson; Meredith, Carole; and Fisher, Kathleen. The
      Science of Scientific Writing. Dubuque, Iowa: Kendall/
      Hunt, 1977.   111p.

This textbook is geared to students in the sciences, and the emphasis is on "data management and organization rather than on style or grammar. " Chapter 1 outlines the approach to scientific writing based on three concepts: "The Relationship Between Thinking and Writing, " "The Relationship Between Structure and Meaning, " and "Reading and Writing as Conditioned Behaviors. " Chapters 2-9 cover such topics as planning for writing, organization, visual aids, reader feedback, formats, style differences in scientific writing, standards for judging one's own work, and proofreading. Chapter 10, the final chapter, summarizes the systematic approach of the book. Most chapters end in exercises, with answers at the end of the book. A glossary of terms used in the book is included.

550   Morrin, Helen C. Communication for Nurses. 1959; rpt.
      Totowa, N. J. :   Littlefield, Adams, 1961. Index. Bib.
      194p.

This text is intended as a reference for beginning nursing students. The 11 chapters cover such topics as "Basic Grammar, " "Sources of Information, " "Writing, " "Speaking, " and "Patient Care Studies. " The chapters end in practice exercises. The appendix is a list of prefixes and suffixes and abbreviations designed to help the student understand scientific words. There is a 71-page glossary of medical terms.

551   Morris, Jackson E. Principles of Scientific and Technical
      Writing. New York: McGraw-Hill, 1966. Index. 257p.

The 13 chapters in this textbook cover such topics as sentence structure, the history of technical writing and types of reports, documents, and papers used in industry. There is a glossary of grammatical terms. The two appendixes cover grammar terms and diagrams of word groups.

552   Morris, John O. Make Yourself Clear! New York: McGraw-
      Hill, 1972. Index. 226p.

This book is based on the author's communication workshops. There are three major sections: (1) the purpose of a particular communication, the audience for the communication, and the needs of the audience; (2) structure of parts, appropriate sequence opening and closing; (3) means and habits of writing, reading, and listening effectively.

This book is intended as a self-improvement course for people in various levels of management.   The author stresses close observance of his five BRISLEDITCH guides: an acronym for "be BRIef ... be SimpLE ... be DIrecT ... be Clear ... be Human."

553 Morris, Richard H.   Credit and Collection Letters: New Techniques to Make Them Work.   Great Neck, N.Y.:   Channel, 1960.   295p.
The first three chapters of this book give practical advice on credit and collection letters, the psychology of customers, and the way to handle delicate situations.   Chapters 4-10 cover effective openings, closings, clarity, explicitness, and individual style.   Chapters 11-18 deal with the process of collection and contain a large number of model letters.   Chapters 19-23 cover credit letters--how to be diplomatic, how to give or request financial information, how to handle orders promptly, how to allow or refuse cash discounts, and how to grant or refuse extensions.   Chapters 24-26 give sound advice on letters that save money for the company, increase sales, build goodwill, or go to foreign countries.   Chapter 27 lists in question-and-answer form what secretaries and bosses should know about letter mechanics.

554 Morrison, Robert H., and Montgomery, Josephine.   Profit-Making Letters for Hotels and Restaurants.   New York: Ahrens, 1959.   180p.
The 18 chapters of this book contain advice for managers, including "Adapting to the Individual," "Starting and Stopping," "Dear Mr. Good Will," "Direct Mail Advertising and Sales Letters," "Credit Letters," and "Routine Letter Jobs."   Chapter 18, "Legal Aspects of Hotel and Restaurant Correspondence," discusses contract offers made in letters, accepting offers, libel, extortion, and credit.   There are numerous examples from hotels and restaurants across the country.

555 Morrison, Robert H., and Sundberg, Trudy.   Bank Correspondence Handbook.   Boston: Bankers, 1964.   236p.
The 20 chapters in this handbook are divided into four parts.   Part I, "The Fundamentals," deals with effective style, openings and closings, and attention getting.   Part II, "Kinds of Letters," includes foreign correspondence.   Part III, "Building and Loan Associations, Investment Companies, Brokerage Firms," deals with correspondence for other financial institutions.   Part IV, "Reference Section," discusses the legal aspects of bank correspondence, letter format, and general principles of style and taste.   Part V, "Some Sample Letters," is a collection of letters illustrating the types discussed in Part II.   There are two appendixes: a list of misused words and a punctuation guide used by the Detroit Bank of Detroit, Michigan.

556 Morton, L. T., ed.   Use of Medical Literature.   2nd ed.   London and Boston: Butterworths, 1977.

557 Moser, Robert H., and DiCyan, Erwin.   Adventures in Medical Writing.   Springfield, Ill.:   Thomas, 1970.   67p.
This book is composed of six essays on writing problems, each essay

by someone in the medical field. Subjects range from instruction on how to use personal experience for enlivening prose to advice on how to avoid overly technical jargon.

558 Mountford, Alan. English in Agriculture. London: Oxford University Press, n. d.

559 Mountford, Alan. English in Workshop Practice. London: Oxford University Press, 1975. 146p.
This textbook for non-native speakers of English is divided into eight units. The units deal with such topics as calipers, bench work, sheet-metal work, and the forge. Each unit begins with short passages and comprehension questions. Following are exercises calling for sentence writing or writing assignments. Answers to the comprehension questions are included. Unit 8 provides four long passages followed by exercises.

560 Moyer, Ruth. Business English Basics. New York: Wiley, 1980.

561 Mullins, Carolyn J. A Guide to Writing and Publishing in the Social and Behavioral Sciences. New York: Wiley, 1977. Index. Bib. 447p.
Addressed primarily to scientists, this book is also a guide for editors of scientific material. Part I, "Outlines, First Drafts, Revisions, and Resources," covers preparation of the manuscript, including advice on avoiding sexist language, usage, handling criticism from others, and coauthors. Part II concentrates on journal articles, including a tabulation of the content and readership of 540 journals in the behavioral and social sciences. One chapter covers the process of publishing from the editor's view, with information on publication lags, rejection rates, and costs. Part III, "General Instructions for Preparing a Book Manuscript," and Part IV, "Publishers, Prospectuses, and Contracts," deal with the special problems of publishing textbooks and edited collections. There are chapters on choosing a book publisher and preparing a prospectus. Included is a matrix table on "Where to Find Information in This Book" and an outline at the beginning of each chapter. Paragraphs and examples are numbered and cross-referenced.

562 Murdock, Michael L. Effective Writing for Business and Government. 2nd ed. Washington, D. C.: Transematics, 1978. Index. 116p.
This guide of eight chapters is intended for someone on the job. Chapters 1-3 cover the writing process; Chapters 4-7, grammar, usage, punctuation, and spelling; and Chapter 8, such points as abbreviations, idioms, possessives, and tense. The appendix contains 11 rewriting exercises based on material in the chapters.

563 Murphy, Dennis. Better Business Communication. New York: McGraw-Hill, 1957. Index. 306p.
This book covers the basic elements of business communication in 17 chapters. Topics include logical thinking, creative thinking,

dictating, public speaking, sentences, letters, and reports. Chapter 1 stresses the need for good communication, and Chapter 17 discusses what management can do to improve communication. Each chapter ends in three or four problems for the reader to consider.

564 Murphy, Herta A. , and Peck, Charles E. Effective Business Communication. 3rd ed. New York: McGraw-Hill, 1980. Index. 741p.
The 19 chapters are divided among five parts. Part I, "Background for Communicating," covers the theory and mechanics of communicating in business. Part II, "Major Letter Plans," covers the purpose of messages, such as direct requests and good and bad news. Part III, "Specialized Messages," includes job applications and collections. Part IV covers short and formal reports. Part V treats group and interpersonal/oral communication. Three appendixes provide the legal aspects of business communication, a 21-page handbook of mechanics and style, and a correction chart for marking papers. Numerous checklists are included.

565 Murphy, Karl. Modern Business Letters. New York: Houghton Mifflin, n.d.

566 Murray, Melba W. Engineered Report Writing. rev. ed. Tulsa, Okla.: Petroleum, 1969. Bib. 121p.
This book, aimed at those on the job, presents the process of writing as analogous to the process of solving an engineering problem, that is, requiring analysis and planning. There are no numbered chapters or sections, but the main units include "Problem Analysis," "How to Plan and Structure a Functional Report," and "How to Edit for Clear Writing." Appendix A is a questionnaire about the intended reader of a report; Appendix B is a questionnaire about the content of a report. Appendix C discusses ways to set up tables of comparisons. Appendix D is a "Verb Test" consisting of four tests with fill-in questions. Appendix E gives the answers to the tests.

567 Naether, Carl Albert. The Business Letter: Its Principles and Problems. New York: Appleton, 1923. Index. Bib. 516p.
This textbook has 11 chapters divided into two parts. Part I, "Essentials of the Business Letter," is composed of three chapters that cover the mission, characteristics, and "dress" of the "modern" business letter. Part II, "Routine and Sales Letters," applies the principles of Part I to various types: order, adjustment, credit, collection, sales, etc. The appendixes contain a three-item bibliography, an outline for analyzing sales follow-up letters, and subjects for student business reports.

568 Nauheim, Ferd. Business Letters That Turn Inquiries into Sales. Englewood Cliffs, N.J.: Prentice-Hall, 1957. Index. 240p.
This book is concerned with goodwill and the "you attitude" in letters to customers and potential customers. The underlying principle is that a letter's purpose is always to sell as well as to handle the

primary problem.   The 16 chapters cover such topics as "How to Capitalize on a 'Yes,'" "How to Say 'No' with a Smile," "How to Make Your Letters Win Cooperation," "How to Handle Responses to Advertising," and "How Letters of Reply Are Followed Up." Chapter 16, "How to Handle Delicate Situations," contains six problem situations that require letters.   Readers are encouraged to write a response and then compare their answers with the sample replies given by the author.   The sample letters are followed by explanatory analysis of the rationale for the approach in the letter.   Each chapter ends with a checklist of important points.

569   Nauheim, Ferd.   Salesman's Complete Model Letter Handbook.
        Englewood Cliffs, N. J. :   Prentice-Hall, 1967.

570   Neelameghan, A.   Presentation of Ideas in Technical Writing.
        Delhi, India: Vikas, 1975.   Index.   Bib.   189p.
This book is divided into 20 sections.   The first 18 discuss the use of symbols in communication, the systematic character and levels of technical writing, organization and format, the nature and properties of ideas, methods of organization, language, numbers, and the law of parameter.   The last two sections are the appendixes.   Appendix I discusses the role of seminal mnemonics in denoting equivalent ideas in a technical document.   Appendix II gives examples of the use of seminal mnemonics.   The book provides an elaborate epistomological analysis of technical writing based on the mystic tradition of Chaldea and India.

571   Nelson, John G.   Preliminary Investigation and Police Report-
        ing: A Complete Guide to Police Written Communication.
        Glencoe Press Criminal Justice Series.   Beverly Hills,
        Calif. :   Glencoe, 1970.   Index.   513p.
This textbook deals with detailed definitions of various types of arrests and crimes.   The author treats reporting and investigating together because they are part of each other in police work.   Directed toward students and professionals, the book has 196 lessons, "each of which can be taught in a period of about fifteen minutes."   Part I includes preliminary investigation and report writing.   Part II is "The Lawman's English Guide," a complete handbook of grammar, punctuation, spelling, and pronunciation.   Parts III and IV cover general skills in investigating and reporting crimes.   There are field problems for class discussion and a glossary of police terms.

572   Nelson, Joseph Raleigh.   Writing the Technical Report.   3rd
        ed.   New York: McGraw-Hill, 1952.   Index.   356p.
Originally published in 1940, this textbook is aimed at senior and graduate students in engineering.   The emphasis is on the principles of designing and organizing reports rather than on the principles of sentences and paragraphs.   The book does cover paragraph coherence and organization.   The five parts cover ways to help the reader, the mechanics of the report, sample reports to criticize, and introductions.   Case problems are included.

573   Newcomb, Duane G.   Word Power Makes the Difference: Making

What You Write Pay Off.   West Nyack, N.Y.:   Parker,
1975.   204p.
This book is designed to guide those in business in writing clear and
effective memos primarily through word choice.   Among other topics
in the 12 chapters are motivation by thinking of the reader's needs,
opening a letter or report, and readability.   The first four chapters
are aimed at getting the writing more readable; the second four con-
centrate on achieving emphasis, and the last four cover motivating
the reader.   The book provides examples and tables for evaluating
the effectiveness of writing.

574   Newcomb, Robert.   Developing Effective Supervisory Newsletters.
      New York: American Management Association,   1956.
      Bib.   83p.
This book is an eight-chapter how-to-do-it guide for producing newsletters: writing, planning, format, distribution, and illustration.
Chapters 1 and 2 provide an overview of the newsletter; Chapter 3
covers topics generally contained in newsletters and the general responsibilities of the editor.   Chapter 4 presents a case study of a
newsletter and shows how management got the participation of its
supervisory personnel.   Chapter 5 covers good writing and appropriate formats.   Chapter 6 contains over 30 sample newsletters with
comments about each.   Chapter 7 covers production and distribution,
and Chapter 8 provides suggestions and material for evaluation of
the newsletter effectiveness.

575   Newman, Bernard H., and Oliverio, Mary E.   Business Com-
      munications: A Managerial Approach.   Pittsburgh: Mo-
      nongahela, 1976.   Index.   212p.
The introduction to this book states that it is intended for those business people and students who want a guide to business writing in
English, and the "English of this book is the English of contemporary
business in the United States."   There are no exercises, but there
is a "Supplementary Guide" for students and a "brief manual" for
instructors.   The 20 chapters are in four parts: "Prerequisites to
Effective Communications," "Written Communications," "Oral Communications," and "Grammar Review."   Appendix A is a list of irregular verbs, and Appendix B is a list of colloquialisms and their
translations, such as "out of kilter" interpreted as "disorderly, in
disorder, not functioning efficiently."   The model letters use European as well as American names and addresses.   The language in
the model letters is somewhat formal: "Your balance is now in
part appreciably past due."

576   Nichols, J. L.   The Business Guide or Safe Methods of Busi-
      ness.   Naperville, Ill.:   Nichols, 1910.   Index.   439p.
Originally published in 1886, this general guide for business practices contains sections on business spelling, penmanship, correspondence, commercial forms, and legal forms.   It also has sections
on parliamentary rules and a final section on business abbreviations
and terms and a legal dictionary.   Among the types of letters discussed are order letters, collection letters, letters of application
and recommendation, and personal letters in business.   The author

assumes that letters will be handwritten and points out that "the essential qualities of a business letter are clearness, neatness and
brevity." Although the author asserts that "flourishing of penmanship or language is out of place in a business letter," the examples
contain rather ornate sentences compared with modern style.

577 Noland, Robert L. Research and Report Writing in the Behav-
         ioral Sciences. Springfield, Ill.: Thomas, 1970. Index.
         Bib. 98p.
This book is a "highly detailed yet practical guide to effective library
research and report writing. It is intended primarily for college
undergraduates, though it may be ... of value to the beginning graduate level also." The focus is on psychiatry, psychology, sociology,
educational psychology, cultural anthropology, and managerial psychology. Chapter 1 is an introduction to report writing and research
methodology in the behavioral sciences. Chapter 2 covers the literature search; Chapter 3, note taking through preliminary writing and
organizing; and Chapter 4, format, giving samples of projects. There
are four appendixes: a bibliography of material on research problems and techniques; a bibliography of report writing; a glossary of
behavioral-science abbreviations; and a list of standard English-Latin
abbreviations.

578 Norgaard, Margaret. A Technical Writer's Handbook for Tech-
         nicians, Engineers, Educators, Businessmen, and Scien-
         tists. New York: Harper and Brothers, 1959. Index.
         Bib. 241p.
Some of this book's 11 chapters, such as "Punctuation," "Abbreviations," and "Grammar," serve exclusively as reference sections.
Other chapters, such as "The Writer and the Reader," "Words and
Their Meanings," and "Writing Procedures," serve as narrative explanations of principles. Following the chapters are two appendixes:
one contains articles from writers in science and technology, and
the other contains a brief list of reference books. The articles are
especially useful in illustrating well-known scientists as good writers.

579 O'Connor, Andrea. Writing for Nursing Publications. Thoro-
         fare, N.J.: Slack, 1976.

580 O'Connor, Maeve, and Woodford, F. Peter. Writing Scientific
         Papers in English. Amsterdam, Netherlands: Associated
         Scientific Publishers, 1975. Index. Bib. 108p.
Commissioned by the European Association of Editors of Biological
Periodicals, the book does offer some grammar advice particularly
appropriate to European scientists who are non-native English speakers. It does not, however, attempt to teach English. The organization of the nine chapters follows the production of a scientific paper.
Chapter 1, "Planning," discusses questions preliminary to writing a
paper: what to submit, when, and where. Chapters 2-4 cover
prewriting, the first draft, and revision. Chapter 5 reviews indexing and inserting final references. Chapter 6 is on the mechanics
of typing, and Chapter 7 provides a checklist for submission of the
paper. Chapter 8, "Responding to the Editor," is a discussion of

advice about what to do if the editor wants changes or rejects the
paper.   Chapter 9 discusses copy editing.   Five appendixes cover
usage and style.

581   O'Hayre, John.   Gobbledygook Has Gotta Go.   Washington,
          D. C.:   U. S. Government Printing Office, 1966.   113p.
This book is a collection of short essays by the author and humorous
illustrations that illustrate the problems of gobbledygook and show
how to avoid the problems.   Some of the 16 sections cover a "Weird
Way of Abstraction, " "High Cost of the Written Word, " and "News
Release Writing--Mostly About Leads. "

582   Oliu, Walter E. ; Brusaw, Charles T. ; and Alred, Gerald J.
          Writing That Works: How to Write Effectively on the
          Job.   New York: St. Martin's, 1980.   Index.   446p.
This textbook, designed for occupational students, has 17 chapters
and a 76-page handbook of grammar and punctuation.   The first three
chapters cover preparation, writing, and revising.   Chapters 4-5
review organization and methods of development.   Chapter 6 treats
effective sentences, and Chapter 7, revising for precision and conciseness.   Chapters 8-9 deal with spelling and vocabulary building.
Chapters 10-14 cover business correspondence, informal reports, research methods, formal reports, and written forms, such as proposals, minutes, and business forms.   Chapters 15-17 discuss tables
and illustrations, oral presentations, and finding a job.   All chapters
end with summaries and writing exercises.

583   Oliver, Leslie M.   Technical Exposition: A Textbook for
          Courses in Expository Writing for Students of Engineer-
          ing.   New York: McGraw-Hill, 1940.   Index.   193p.
The 13 chapters in this textbook cover the qualities of good engineering writing, effective sentence structure and style, revision, types
of writing tasks, reports, research papers, organization, and manuscript form.   Chapters 10-11 discuss the business letter.   Chapter
12 advises on increasing vocabulary, and Chapter 13 gives rules of
punctuation.   Review questions are included.

584   Opydycke, J. B. , and Drew, C. A.   Commercial Letters.
          New York: Holt, 1918.

585   Opdycke, John B.   Get It Right!   rev. ed.   New York: Funk
          and Wagnalls,   1941.   Index.   673p.
This book is a general reference on writing style.   The 20 chapters
include treatments of direct mail, letter writing, reports, and telegrams.   Other chapters cover such topics as journalism and advertising.   The final handbook section reviews style, grammar, and
punctuation.

586   Opdycke, John B.   Take a Letter Please!   New York: Funk
          and Wagnalls,   1944.   Index.   479p.
The aim of this book is to guide people in the proper style of writing
letters for friendly effect and to express personality.   Chapter 1,
"The Quick and the Dead, " discusses the quality of life in letters

and gives advice on social letters of apology, appreciation, condolence, and the like.   Chapter 2, "The Yea and the Nay," deals with eliminating clichés and achieving clarity and conciseness.   The examples in Chapters 1 and 2 are taken from such greats of English literature as Samuel Johnson and Daniel Defoe.   Chapter 3, "The Frame and the Picture," discusses format.   Chapters 4-9 cover types of letters, such as application, inquiries, adjustment, and collection.   The three appendixes deal with the post office and the law.   "The Dead Letter Office" explains the operation of that office. "The Letter and the Law" explains penalties for sending obscene or illegal items through the mails.   "The Letter and the Courts" explains how business transactions are handled if letters are lost in the mail and the validity of contracts made through the mail.

587   Orcutt, W. D.   Writer's Desk Book.   New York: Stokes, 1912.

588   Orientation in Business English.   Silver Springs, Md. :   Institute of Modern Languages, n. d.

589   Orlich, Donald C. , and Orlich, Patricia Rend.   The Art of Writing Successful R & D Proposals.   Pleasantville, N. Y. :   Redgrave, 1977.   Index.   73p.
The stress of this book is on federal funding.   The six chapters are "Organizing Your Ideas," "Writing the Proposal," "Selecting Appropriate Research Designs," "Project Evaluation," "Other Proposal Components," and "Submitting a Proposal."   Chapter 1 contains sample abstracts.   The book has headings in the extra wide margins.

590   Otte, Frank R.   Complete Book of Extraordinary Collection Letters.   Englewood Cliffs, N. J. :   Prentice-Hall, 1965.

591   Palen, Jennie M.   Report Writing for Accountants.   Englewood Cliffs, N. J. :   Prentice-Hall, 1955.   Index.   602p.
This book covers writing techniques for accounting reports in five of its 29 chapters.   The remaining chapters cover the technical components of the report.   Chapters 23-25--"The Technique of Writing the Comments I, II, and III"--cover standard expository skills and include advice on hedging and writing preambles.   Chapter 26, "The Right Word," treats the use of jargon, technical terms, and intensifiers.   Chapter 27, "Presentations," gives some advice about format and letter mechanics.

592   Palmer, Herbert H.   Tested Sales Letters.   New York: McGraw-Hill, 1935.   Index.   530p.
This guide to letter writing has 18 chapters covering such topics as sales letters, attracting customers, special events, helping sales people, follow-up letters, complete campaign, format and stationery, costs, and letters to dealers.   The appendix gives sources for mailing-list information.   The author has tried to include sample letters dealing with every phase of a complete sales campaign.   This book is essentially a book of model letters analyzed for their effectiveness.

593 Palmer, O. R. <u>Type-Writing and Business Correspondence.</u>
     Philadelphia: Lippincott, 1892 (also 1893, 1896, 1900,
     1905).

594 Park, Clyde W. <u>English Applied in Technical Writing.</u> New
     York: Crofts, 1926. Index. 313p.
This textbook was developed as a way to fill the needs of the "average technical student." The 12 chapters include "The Point of View,"
"Good Mechanical Form," "Logical Organization of Material," "Clear
Statement of Ideas," and "Accurate Use of Words." There are numerous illustrations throughout the chapters. Chapter 11, "The
Technical Writer's Literary Background," stresses the importance
of good reading for expressing ideas--rejecting the idea of a "narrow" view for the technical writer. The chapter includes "An Engineer's Private Reading List" from the March 1925 edition of <u>West-
inghouse Electric News.</u> This list emphasizes masculine adventure
and science fiction. A supplementary list has a wider range (Jane
Austen to Owen Wister). Chapter 12, "Suggested Exercises," is for
classroom use and includes an exercise of recopying one of the illustrations of a specification, a practice in abbreviations, and rewriting or developing paragraphs. There are exercises based on
the literary reading that require creative writing or literary criticism.

595 Parkhurst, Charles C. <u>Business Communication for Better
     Human Relations.</u> 7th ed. Englewood Cliffs, N. J.:
     Prentice-Hall, 1966. Index. 519p.
This textbook is divided into three parts: "Fundamentals of Effective
Communication," "Types of Business Communications," and "Reference Sections." The 12 chapters in the first two parts cover collections, sales letters, planning and writing the letter, and customer-
service letters. The six chapters in Part III cover grammar, punctuation, usage, forms of addresses, and the telephone and telegram.
Each chapter in Parts I and II ends in discussion questions, writing
exercises, or assignments and cases. Exercises and problems come
from company files--a long list of contributors appears in the front
of the book. About 40 case studies and many illustrations of letter
forms and types are included.

596 Parkhurst, Charles C. <u>Case Studies and Problems in Business
     Communication.</u> Englewood Cliffs, N. J.: Prentice-Hall,
     1960. 140p.
This five-part book deals with sample situations requiring written
communication. Conditions and circumstances of each problem are
given, and a preliminary class discussion is advised. Part I requires letters with a personal tone, conveying news of hiring, dis-
missal, condolence, and congratulation. Part II covers credit and
collection situations, and Part III, with maintaining good public relations with complaint and adjustment letters. Part IV deals with
promotional and sales literature. Part V gives assignments relative
to compiling surveys, questionnaires, and reports.

597 Parkhurst, Charles C. <u>English for Business.</u> 4th ed. En-
     glewood Cliffs, N. J.: Prentice-Hall, 1963. Index. 423p.

This textbook has 18 chapters divided in two parts. Part I covers grammar, punctuation, usage, and the format of business letters. Part II illustrates types of business letters, including collection, sales, and application. Chapters are divided into sections, each followed by discussion questions, case problems, and a vocabulary list. Also included is a list of 450 commonly misspelled words and a checklist for good business letters.

598 Parkhurst, Charles C.   Modern Executive's Guide to Effective Communication. Englewood Cliffs, N. J. : Prentice-Hall, 1962. Index. 535p.
In this book of 12 chapters, the author stresses the need for good public relations in business communications. Part I covers communication and human relations, planning and writing letters, image improvement, and letter format. Part II covers various types of business communications: inquiries, orders, responses, the business report, and business and personal, public-relations, customer-service, sales and promotional, credit, and collection letters. Part III serves as a 96-page handbook of grammar, punctuation, commonly misused words, and address forms.

599 Parkhurst, Charles C.   Practical Problems in English for Business. Englewood Cliffs, N. J. : Prentice-Hall, 1963.

600 Parkinson, Joy. English for Doctors and Nurses. London: Evans Brothers, 1978. 124p.
This handbook is designed as a review of grammar and usage for those who are not native speakers of English. There are 44 units, each covering one grammatical or usage point. Each unit is introduced by a short dialogue or conversational paragraph illustrating the principle or explaining it. Topics include "Adjectives," "Some and Any," "Verbs and Tenses," "Interrogative Forms," "Prepositions Followed by the Gerund," "Had Better," and "Lie and Lay." There are no exercises; there is no table of contents or index; and the units are not in alphabetical order or grouped by type.

601 Parr, William M. Executive's Guide to Effective Letters and Reports. West Nyack, N. Y. : Parker, 1976.

602 Passman, Sidney. Scientific and Technological Communication. Oxford and New York: Pergamon, 1969. Index. Bib. 151p.
The purpose of this work is to assess the causes of the "information explosion" and offer some "ameliorative action in this field [scientific communication]." The nine chapters deal with information transfer, report writing, language, and research as they relate to the communication process. Chapter 9 focuses on the international aspects of scientific and technical communication--the language gap and international coordination of communication. The appendix includes a taxonomy of eight types of technical reports and articles.

603 Patterson, Frank M. Police Report Writing for In-Service Officers. Springfield, Ill. : Thomas, 1977. Index. 144p.

This textbook is for the experienced police officer.   Chapters 1-3
deal with general principles of writing (grammar and style).   Chapters 4-9 deal with reports on such offenses as burglary, assault,
and robbery.   The last three chapters treat the writing of memos,
correspondence, and recommendations.   Each of the 12 chapters has
writing and discussion assignments.

604   Patterson, Frank M. , and Smith, Patrick D.   A Manual of
          Police Report Writing.   Springfield, Ill. :   Thomas, 1968.
          Index.   78p.
According to its author, this "manual is intended to provide instruction in the composition of police narrative writing primarily in terms
of the organization and the language of such reports.   No treatment
has been given to the various kinds of reports that police personnel
are required to fill out.... "   These forms are ignored because
"there is little agreement" on what such reports should include and
they are written by "supervisory personnel who are already experienced writers. "   This book aims to "help the beginning policeman. "
Part I, "Definition of a Report, " covers the purpose of the police
report.   Part II, "The Organization of a Report, " covers the criteria
for including information, the logic used and organization, paragraphing, unity, coherence, and description.   Part III, "The Language of
Reports, " treats language history, usage, wordiness, denotation and
connotation, judgment, hearsay, and stated opinion.   Part IV, "Aids
in Writing Reports, " gives advice on using supplementary books, a
list of commonly misspelled words, and dictating tips.   The final
section of Part IV provides proofreading exercises for a police-
report writing course.

605   Pauley, Steven E.   Technical Report Writing Today.   2nd ed.
          Boston: Houghton Mifflin, 1979.   Index.   321p.
This textbook suggests that students "write papers about their technical areas and aim their writing at an uninformed reader. "   The
book's 17 chapters, in five parts, cover such topics as interpreting
statistics, oral reports, defining and describing a mechanism, and
business letters.   Included are writing assignments, exercises, and
student examples.

606   Pearsall, Thomas E.   Audience Analysis for Technical Writing.
          Beverly Hills, Calif. : Glencoe, 1969.   113p.
Designed as a supplement to technical writing textbooks, this book
aims to help students better adapt their writing to the needs of their
readers.   In the introductory section, "Audience Analysis, " the author defines five types of audiences: The Layman, The Executive,
The Expert, The Technician, and The Operator.   The author emphasizes that, if the students know the needs of these audiences,
they are better able to select the content, adapt the style, and determine the technical background appropriate to the reader.   The
rest of the book is divided into two parts, "Undersea Exploration"
and "Space Exploration. "   Both parts contain five full writing samples, each aimed at one of the five audiences.

607   Pearsall, Thomas E.   Teaching Technical Writing: Methods

for College English Teachers. Washington, D.C.: Society for Technical Communication, 1975. Bib. 23p.
Designed for English teachers who are teaching technical writing for the first time, this book begins with a definition and discussion of technical writing style and pedagogical approaches. It then discusses how to make writing assignments and classroom activities. It also discusses typical ways of arranging a course, although no course syllabus is provided. It concludes with a bibliography of journals available, the major bibliographies, and articles on pedagogy.

608   Pearsall, Thomas E., and Cunningham, Donald H. How to Write for the World of Work. New York: Holt, Rinehart and Winston, 1978. Index. Bib. 369p.
This textbook is designed for "the student who is being educated for a specific vocation or profession ... in a vocational or technical institute, in a two-year college, or in a pre-professional university program." The 14 chapters are divided into two parts: Unit I, "Correspondence," and Unit II, "Reports." Following the two units are four "annexes": a "Writer's Guide" for grammar and mechanics, "Formal Elements of Reports," "Library Research," and "Metric Conversion Tables." Following these sections is a selected bibliography. Among the special topics covered are accident reports, bibliographies, and literature reviews, and proposals (solicited and unsolicited). The authors provide "suggestions for applying your knowledge" (short sections of writing assignments and other related projects).

609   Peirce, J. F. Organization & Outlining: How to Develop & Prepare Papers, Reports, & Speeches. New York: Arco, 1971. 79p.
This book on organization and outlining was an outgrowth of a course the author gave for the Federal Systems Division of IBM, Houston. The author says the value of the book lies in the fact that it "contains detailed discussions of twenty-two patterns of organization and one or more examples of each as well as cross references to related patterns and examples." The five chapters discuss the importance of organizing, types of outlines, outline form and evaluation, principles and patterns of organization, and development. There are 45 examples and checklists of patterns of organization.

610   Perry, Sherman. Making Letters Talk Business. Middletown, Ohio: American Rolling Mill, 1924. Index. 206p.
This book is divided into four sections, each with its own table of contents. The first section, "Essential Principles," discusses such topics as clarity, conciseness, persuasion, tact, sales messages, transition, openings and closings, and variety in word choice. "Essential Mechanics" is addressed primarily to the stenographer, covering punctuation, use of figures, accuracy versus speed, salutations, spacing, etc. "Words Frequently Misspelled" is a 38-page list of words with their Gregg Shorthand symbols. "Essential English" covers parts of speech, usage, and sentence structure. Included is a discussion of "Where the Stenographer Fails." "Essentials of Report Writing" reviews gathering data, organization, and format.

The final chapter, "The Way It's Written, " is a copy of the author's talk in April 24, 1924, at the College of Commerce at the University of Illinois.   He states, "Although we have heard much discussion and criticism of the crudeness of the business man's English, I can assure you there are many business men who have the highest regard for the written or spoken word.... In other words, the day has come when executives are asking: 'Can you write and speak clearly, forcibly, precisely?'"

611   Persing, Bobbye Sorrels.   The Nonsexist Communicator.   East Elmhurst, N.Y.:   Communication Dynamics, 1978.

612   Personal Side of Writing Letters in Business.   Englewood Cliffs, N.J.:   Prentice-Hall, 1967.

613   Peterson, Martin S.   Scientific Thinking and Scientific Writing. New York: Reinhold, 1961.   Index.   Bib.   215p. Addressed to students in the biological and physical sciences, this book links scientific writing to scientific logic, based on organization.   Chapter 1, "The Genesis of Scientific Thought and Writing, " illustrates ways in which thought relates to writing--e.g. , inductive thinking becomes an inductive paragraph.   Chapter 2 relates the classical methods of scientific reasoning to the organization of scientific articles and reports.   Chapter 3 discusses experimental design and interpretation of data as they relate to statements in articles and reports.   Chapter 4 discusses ways of developing a systematic approach to both scientific thinking and writing and provides an annotated list of works related to this subject.   Chapter 5 gives advice and shows examples of various types of scientific writing--e.g. , book review, essay, talk.   Although there are no exercises for students, there are chapter summaries.

614   Phillips, Bonnie D.   Effective Business Communications.   New York: Van Nostrand Reinhold, 1977.   Index.   247p. This textbook is aimed at two-year colleges or technical schools. The 13 chapters are divided into four sections: "Listening and Speaking, " "Background for Business Writing, " "Letter Writing, " and "Report Writing. "   Chapters include grammar and usage, writing mechanics, form messages, persuasive messages, reporting numerical data, and report planning and writing.   The oral-communication chapter covers telephone technique and dictating.   There are exercises interspersed throughout the chapters.   Numerous examples of business communications are offered.   The appendix includes a guide to using numbers, lists of difficult words, abbreviations, typing styles, and forms of address.

615   Picken, James H. , ed.   Business Correspondence Handbook. Chicago and New York: Shaw, 1927.   Index.   Bib.   836p. This book is "a reference work for business men who write and use letters, and ... a source from which the student may glean letter-writing information.... "   The 30 chapters of this book cover such subjects as planning letters, various types of letters, how to analyze a business prospect, the psychology of effective letter writing, and

direct-mail advertising.   The appendix contains a glossary of business correspondence terminology and a bibliography.

616   Pickett, Nell Ann.   Business Letters.   New York: Harper and
        Row, 1975.

617   Pickett, Nell Ann.   Practical Communication.   New York:
        Harper's College Press, 1975.   Index.   259p.
This textbook has 26 chapters arranged in seven parts: "Developing
the Communication: Analysis, " "Developing the Communication: Description, " "Writing Reports, " "Writing Business Letters, " "Using the
Library, " "Using Visual Materials in Written Communication, " and
"Developing Oral Communication. "   Writing assignments are interspersed through the chapters.

618   Pickett, Nell Ann, and Laster, Ann A.   Technical English:
        Writing, Reading and Speaking.   3rd ed.   San Francisco:
        Canfield, 1980.   Index.   627p.
This textbook (formerly titled Writing and Reading in Technical English) is designed for technical students in community colleges and
technical schools.   The book is divided into three parts.   Part I,
"Forms of Communication, " comprises three-fourths of the book and
covers such topics as processes, instructions, descriptions, analysis, business letters, the library paper, reports, oral communication, and visual aids.   Each of the chapters in Part I begins with a
list of behavioral objectives, contains student plan sheets, and ends
with exercises.   The pages are perforated so that the instructor can
also use the text as a workbook.   Part II, "Readings, " opens with
a brief discussion of reading skills and contains selected readings
for classroom discussion.   Part III, "Handbook, " treats paragraphing, mechanics, and grammatical usage.

619   Piper, Henry D. , and Davie, Frank E.   Guide to Technical
        Reports.   New York: Rinehart, 1958.   Bib.   83p.
This work grew out of an industrial style manual for the Shell Oil
Company, known as the "Shell Guide. "   Essentially, the book is
an expanded outline with three sections.   Section I, "Preliminary
Considerations, " covers "fundamentals of the report" (such as organization, reader's point of view, emphasis through format, and
tactful presentation), "method of writing the report" (preliminaries,
outline, first draft, and revision), and "types of reports" (formal
and informal).   Section II, "Standard Procedures for Preparing Reports, " covers parts of formal and informal reports and the treatment of "other elements" (such as abbreviations and illustrations).
Section III is "Punctuation, Grammar, and Style. "   The appendix
provides five examples of formal and informal reports, edited with
commentary.

620   Pittman, G. A.   Preparatory Technical English.   New York:
        Longman, 1960.   175p.
This book, written by the Director of the English Language Institute
at Victoria University in Wellington, New Zealand, is aimed at "ap-
prentices and students" who wish to improve "their comprehension

of technical literature written in English and their expression in English of their ideas in technical matters." The book is divided into
26 lessons designed to give the non-native speaker better command
of English. Each lesson is followed by an exercise section that may
ask the student to fill in missing sentence parts, correct expressions, and write sentences and short passages. Some of the lessons
are also followed by a reading section. The lessons are often preceded by line drawings of objects (glasses of water, nails, mechanical devices, etc.). Wherever possible, the author avoids grammatical terms, teaching through questions and answers.

621   Poe, Roy W., and Fruehling, Rosemary T.   Business Commu-
         nication: A Problem Solving Approach. 2nd ed.   New
         York: Gregg/McGraw-Hill, 1978.   Index.   Bib.   358p.
This textbook consists of ten parts. Part I gives "Background for
Business Writing." The remaining parts contain four to ten cases
each. The authors provide a brief introduction to the topic and then
analyze memos and letters in light of the topic. For example, in
Part III, Case 3 is "Writing Promotion and Sales Letters." The
problem is one of writing to executives to sell them on renting a
meeting hall. The authors then discuss this kind of sales letter,
give tips on how to proceed, give examples and analyze them. Part
X gives the student 33 situations to respond to as sales manager of
a publishing company. The reference section covers forms of address, business-letter styles, abbreviations, and proofreader's marks.
The preface states that there is a workbook and a set of tapes with
12 lectures on on-the-job business writing available.

622   Pokress, E.   Research and Technical Writing: A Manual for
         Those Interested in Free-Lance Work or Exciting Employ-
         ment Opportunities.   Allenhurst, N.J.: Aurea, 1965.
         66p.
This is a processed folio (mimeographed) available only from the
publisher. The guide is addressed to anyone from a housewife to a
retiree who wishes to try to enter freelance research writing. The
introduction advises contacting local firms or groups and starting by
doing typing. This start will lead to research assignments. Twelve
chapters cover opportunities, furnishing information and locating sup-
ply sources, preparing speeches and lectures, how to use the library,
nonfiction articles, report writing, how to get more assignments,
educational material, scholarly style and academic format, technical
reports and specifications, technical and scientific manuals, and essentials of reproduction methods. The chapter on scholarly style
focuses on the term paper and advises the new writer to get a college degree because "higher pay will follow almost automatically."

623   Poole, George William, and Buzzell, Jonathan John, eds. Let-
         ters That Make Good: A Desk Book for Business Men.
         3rd ed.   Boston: American Business Book, 1915.   In-
         dex.   448p.
This book, originally copyrighted in 1913, is designed for those who
write letters and manage "correspondence departments." It is divided into two parts: Part I, "The Principles of Letter Writing,"

is composed of ten articles by various authors covering planning, data gathering, style, mechanics, and supervising correspondence. Part II, "Specimens and Examples," contains 363 pages of sample letters for every conceivable circumstance. The sample letters literally are dittoed on letterhead stationery.

624   Powell, John Arthur.   How to Write Business Letters.   Chicago: University of Chicago Press, 1925.   Index.   192p.
This book is intended as a guide for the dictator and the secretary. The topics in the eight chapters include "Atmosphere and Personality," "The Makeup of the Sales Letter," "Miscellaneous Letters," "Why Trouble About Rules of English," "Spelling," and "Punctuation." Three appendixes are included: "Hints on the Mechanics of Letter-Writing," "Guide to the Use of Good English" (a usage glossary), and "Converting Notes into Minutes."

625   Proceedings of the Society for Technical Communication.   Washington, D.C.:   Society for Technical Communication, 1953-present.   Pages vary.
These proceedings are collections of papers given at the yearly International Technical Communication Conferences.   Typical topics are education and research, management and development, visual and audiovisual materials, and writing and editing.

626   Proposals and Their Preparation.   Washington, D.C.:   Society for Technical Communication, 1973.

627   Prout, John.   Adjustment Letters Handbook.   New York: Prentice-Hall, 1954.   Index.   288p.
This book is concerned solely with writing adjustment letters to customers and retaining goodwill.   Part I, "How to Write Adjustment Letters," contains seven chapters covering such items as openings and closings, getting the facts, and promptness.   Part II, "How to Handle Specific Adjustment Problems," contains 17 chapters covering mass complaints, the chronic complainer, form letters, chiselers, and how to say no.   Each chapter begins with a quotation from a business executive about the need for good letters replying to customer complaints.   The book contains numerous examples and letters from real companies.

628   Racker, Joseph.   Technical Writing Techniques for Engineers.   Englewood Cliffs, N.J.:   Prentice-Hall, 1960.   234p.
The five chapters of this book include definition of technical writing, audience analysis, word choice, "Technical Illustrations," and the "Preparation of Technical Manuscripts."   Illustrations are abundant. There is a 106-page glossary at the end of the book, covering "Air Force Terms," "Automation Terms," "Computer Terms," "Electrical and Electronic Terms," "Guided Missile Terms," "Radio and Radar Navigation Terms," and "Space Technology and Transistor Terms."

629   Ramsey, Robert E.   Effective House Organs.   Appleton, 1920. Index.   Bib.   361p.
This textbook has 26 chapters divided into two parts.   Part I deals

with the history of house organs, types of house organs, planning, content, and the role they play in company advertising programs. Part II contains 18 chapters, each explaining how a particular institution or industry uses house organs successfully, e. g. , "How Banks Have Used House Organs Successfully, " "How House Organs Have Been Used in the Apparel Field Successfully. " The chapters cover extensively the house organs in each field. Each chapter ends in several discussion questions or writing exercises. There are six appendixes reviewing the typical costs of house organs, testimonials from companies on the effectiveness of house organs, a discussion of a particularly successful house organ--"The Houghton Line, " a list of 19 ways, with testimonials, that the house organ helps to distribute company products (contests, reminder advertising, reaching out-of-the-way places), the number of house organs published in selected industries, and a discussion of how to sell the idea of a house organ to management. There are numerous illustrations of early house organs.

630 Ranous, Charles A. Communication for Engineers. Boston: Allyn and Bacon, 1964.

631 Ranous, Charles A. The Engineer's Interfaces: A College Course in Effective Communication. Madison, Wis.: Xer-Lith, 1974.

632 Rathbone, Robert R. Communicating Technical Information. Reading, Mass.: Addison-Wesley, 1966. Index. Bib. 104p.
This book is a self-improvement writing guide for engineers and scientists. The ten chapters include "The Tenuous Title, " "The Inadequate Abstract, " "The Wayward Thesis, " "The Neglected Pace, " and "The Arbitrary Editor. " Each chapter has a headnote that is a comment on the craft of writing by writers ranging from Samuel Johnson to an anonymous MIT student. Appendix I is "An Annotated Bibliography" of books on scientific and technical writing. Appendix II is "A Journal Article Before and After Editorial Revision. "

633 Rathbone, Robert R. , and Stone, James B. A Writer's Guide for Engineers and Scientists. Englewood Cliffs, N. J.: Prentice-Hall, 1962. Index. Bib. 348p.
This textbook is designed as a "special source book" that analyzes the inexperienced writer's problems, discusses solutions, and provides a variety of models from actual reports. It is intended for science and engineering students and professionals. Among the topics of special interest are writing introductions; explaining new concepts, methods, and devices; reporting negative results; pace; use of the pronoun "I"; and the writing process. The appendix contains samples of articles and reports and a bibliography of books on report and technical writing. A few exercises and discussion questions follow each of the nine chapters.

634 Rautenstrauch, Walter. Industrial Surveys and Reports. New York: Wiley, 1940. Index. Bib. 189p.
This textbook focuses on the investigation and collection of data and

business policies related to writing reports. Chapter 1, "The General Principles of Report Writing," includes discussion on the importance of reports (economic and otherwise), the "essentials" and "elements" of good reports, and several cases for the student. Chapter 3, "The Contents of the Report," covers the scope of reports and identifies/defines the parts of a report. The remaining five chapters review organizational, financial, and personnel issues in business and industry that may be dealt with in reports. The appendixes include mathematical formulas, a sample report, and a guide for graphic presentation.

635 Raymond, Charles H. Modern Business Writing. New York:
          Century, 1921. Index. 476p.
This book is designed for the businessperson who must design advertising material and write letters. Its 32 chapters are divided into two parts. Part I, "The Selling Appeal," covers advertising techniques and sales letters, including chapters on the "prospect" and the "product" and samples of miscellaneous sales letters and advertisements. Part II, "Everyday Letters," covers inquiry, credit, and adjustment letters; establishing a "house" style; and letter format.

636 Reddick, DeWitt C. Literary Style in Science Writing. New
          York: Magazine Publishers Association, 1969. 42p.
Each of the five chapters of this work covers a single topic designed to help the scientist communicate with the layperson. Emphasis is on science writing for popular magazines and newspapers. Chapter 1, "Lessons from a Pioneer," discusses narration, action verbs, comparison. Chapter 2, "The Nature of Style," covers factors that shape style. Chapter 3, "The Bridge of Comparison," discusses the reasons for using comparisons. Chapter 4, "The Fabric of Narration," gives tips for narrative devices. Chapter 5, "Exposition: A Key to Effective Science Writing," stresses definition, clarity, and logical development.

637 Redmond, Pauline, and Redmond, Wilfred. Business Paper
          Writing--A Career. New York: Pitman, 1939. Index.
          194p.
This text is devoted to training people to write stories for business journals. Each chapter ends in review questions. The ten chapters cover such topics as "Definitions," "How to Find Material," "How to Write a Business Paper Story," "A Complete Tradepaper Story--From Idea to Check." The term "business paper" here is used to mean business trade journal.

638 Reed, Jeanne. Business English. 3rd ed. Continuing Educa-
          tion Series. New York: Gregg/McGraw-Hill, 1978. In-
          dex. 140p.
This unusual package is a boxed packet. It is called a "Gregg Text-Kit for Adult Education." The softcover text has 25 units with practice sentences for students to try after reading the text. Answers are given for student's self-check. The two pads of additional exercises, "Self-Check 1" and "Self-Check 2," give further practice. An answer booklet is for student use. Also included is a survey

to be used before starting so that students can see areas of weak-
ness.   The inside front and back covers of the text contain a checklist of the units so that students can record when they finish a unit
and the percentage of accuracy on the exercises.   All the units deal
with grammar except Unit XXV, which covers letter format, tone,
clarity, and effectiveness.

639   Reid, James M., Jr., and Silleck, Ann.   Better Business
          Letters: A Programmed Book to Develop Skill in Writing.
          2nd ed.   Reading, Mass.: Addison-Wesley, 1978.   203p.
Designed for in-house training programs, this book has six lessons:
"Writing Concise Sentences," "Keeping Your Reader's Interest,"
"Writing for Easy Reading," "Being Natural, Courteous, and Personal," "The Start and the Finish," and "Planning and Writing Your
Letter."   Each lesson has extensive exercises and then a quiz.   Ap-
proximate times for study and taking the quiz are given in "To the
Instructor."   Most lessons with quiz take about 90 minutes.   The
appendix contains letter layout and diagnostic tests.   At the end of
the book are answers for the diagnostic tests and the quizzes.   The
authors advise against doing more than two lessons a day.

640   Reid, James M., Jr., and Wendlinger, Robert M.   Effective
          Letters.   3rd ed.   New York: McGraw-Hill, 1978.

641   Reisman, S. J.   A Style Manual for Technical Writers and
          Editors.   New York: Macmillan, 1962.   Index.   Bib.
          223p.
This handbook of 15 chapters is divided into four parts.   Part I
deals with the functioning of technical-publications departments.   Part
II discusses how to put together reports, proposals, and manuals.
Part III deals with such topics as "Tables," "Illustrations," "Footnotes," "References," "Bibliography," and symbols.   Part IV emphasizes style, usage, sentence structure, and paragraphing.   Appendix A provides specimens of technical writing; Appendix B is a
bibliography; and Appendix C gives "Editorial and Proofreading Marks."

642   Rhodes, Fred H.   Technical Report Writing.   2nd ed.   New
          York: McGraw-Hill, 1961.   Index.   168p.
Although this book "is intended primarily, but not solely, for undergraduate and graduate students in technical curricula," it contains
no exercises or other pedagogical tools.   Evidently, it is designed
as a reference for technical students.   In addition to principles of
good reports, organization, and style, the ten chapters also cover
"The Laboratory Notebook," "Reports as Evidence in Patent Actions,"
and "The Oral Presentation of Technical Reports."   Because of his
concern that "many of those ... engaged in planning and interpreting
experimental work are not acquainted with even the simpler applications of statistical methods," the author includes Chapter 9, on "The
Precision of Results," and Chapter 10, on "Analysis of Correlation
and of Variance."   Three appendixes cover the mathematics of statistical methods.

643   Richards, Jack C., ed.   Teaching English for Science and

Technology. Singapore: Singapore University Press, 1976.

644 Richardson, Lou, and Callahan, Genevieve. The New How to
     Write for Homemakers. 2nd ed. Ames: Iowa State
     University Press, 1962. Index. 201p.
This book is aimed directly at the writing tasks of home economists.
The 21 chapters include "Fresh Look at Communications," "The Topic Sentence," "Recipes," "Menus and Meal Plans," "Booklets, Bulletins, Leaflets," "Educational Films and Slides," "Cookbooks and
Texts," and "Copy Editing and Proofreading." The margins of the
chapters have further writing tips or tips relevant to the chapter's
content. Since home economists must prepare the food for photography as well as write the descriptions, four-color illustrations of
food photography are included. The chapter on writing recipes includes samples of five "patterns," using the same recipe.

645 Rickard, Thomas Arthur. A Guide to Technical Writing. San
     Francisco: Mining and Scientific, 1908. 127p.
This book is of particular historical interest because it is the first
published book on technical writing for the professional. The author
comments, "It has been said that in this age the man of science appears to be only one who has anything to say, and he is the one that
least knows how to say it. ... Write simply and clearly, be accurate and careful; above all, put yourself in the other fellow's place.
Remember the reader." Geared to the mining and metalurgical sciences, the 17 short, unnumbered chapters cover matters of language,
usage, grammar, and mechanics slanted toward the needs of the
technical writer. Included in the discussion of word choice are comments on the use of intensifiers ("Unconsidered Trifles"), avoidance
of journalese (which injects exaggerated opinion into technical writing), and the separation of fact from opinion. The book ends with
a paper the author read before the American Association for the Advancement of Science, at Denver, on August 28, 1901: "A Plea for
Greater Simplicity in the Language of Science." "We must remember," the author suggests, "that language in relation to ideas is a
solvent, the purity and clearness of which effect what it bears in
solution."

646 Riebel, John P. How to Write Reports, Papers, Theses, Ar-
     ticles. 2nd ed. New York: Arco, 1972. 121p. Un-
     paged appendix [54 pages].
This book opens with a copy of a letter to the author from the chief
of the Specifications Section at Douglas Aircraft. The letter discusses the philosophy of Douglas about writing--primarily that it
should be clear and concise. The first major section, "How to
Write Technical Reports, Papers, Articles and Theses," includes
organization, presentation, language, etc. The material is the instructional material used by the author in his course at California
State Polytechnic College in San Luis Obispo. The next section is
made up of unnumbered units on "Grammar," "Punctuation," "Paragraphing," "Special Problems in Composition," "Vocabulary," and
"Letter Writing." The appendix consists of selected readings from
conferences or technical publications.

647   Riebel, John P.   How to Write Successful Business Letters.
            2nd ed.   New York: Arco, 1971.   Index.   276p.
The book is a practical guide and text for a concentrated course of
15 days.   The early chapters deal with audience awareness, writing
plainly, and ending "with a bang."   Later chapters, such as "The
Sunshine of Your Smile" and "A Soft Answer Turneth Away Wrath,"
stress the importance of building goodwill for the business.   The
last three chapters illustrate good business style.   The appendix,
"Modernizing Hackneyed Expressions," is an alphabetical list of alternatives to trite expressions.   There are no exercises; however,
a number of examples of letters with critiques are included.

648   Riebel, John P., and Roberts, Donald R.   Ten Commandments
            for Writing Letters That Get Results.   Pleasantville,
            N. Y.:   Printer's Ink, 1957.   Index.   184p.
This book is divided into "10 C's," called Commandments: "Be
Clear, Be Correct, Be Complete, Be Concise, Be Courteous, Be
Considerate, Be Cheerful, Be Convincing, Be Conversational, Be Clever."   The section "Hall of Fame" contains 30 letters the authors
felt were particularly effective.   The book has numerous examples
of real business letters, and the commentary on the samples forms
the bulk of the book's content.   The final section, "Salutopenings
and Compliendings," suggest openings and closings other than the
traditional ones.   For example, an opening might be, "How are you,
Mr. Boget?"; a closing, "What do you think of this proposal, Mr.
Wiell?"   The index is combined with acknowledgments--all in alphabetical order.   The book begins with a subject index of the types of
letters (credit, sympathy, etc.) found in the book.

649   Roberts, Ffrangcon.   Good English for Medical Writers.   Lon-
            don: Heinemann, 1960.   Index.   Bib.   179p.
This handbook of grammar and usage is intended for the aspiring
medical writer.   The 16 chapters are devoted entirely to usage,
grammar, spelling, technical terms, logic and reasoning, and verbosity.   There is special advice for the practicing professional who
wants to write for medical journals.

650   Roberts, Louise A.   How to Write for Business.   New York:
            Harper and Row, 1978.   Index.   289p.
This textbook has three parts: "The Business Writer's Guide" (grammar, punctuation, and spelling), "The Elements of a Business Style"
(words, sentences, paragraphs, organization, prose passages for
study), and "Practice in Writing Business Form" (short business
forms, long ones, final writing, and reading assignments).   Included
in the chapter on spelling is a usage glossary that also includes a
list of business terms and space for students to write in definitions
and start their own business glossary.   The exercises and readings
are fairly sophisticated.   The last assignment in the book asks the
student to write a modern version of "A Message to Garcia"--an
800-word article on worker motivation offering solutions to the problem of restoring pride in workmanship.

651   Robertson, Horace O., and Carmichael, Vernal H.   Business

Letter English. 2nd ed. New York: Gregg/McGraw-Hill, 1957. Index. 470p.
This textbook emphasizes grammar, mechanics, and usage. Part I, "Business English," devotes 14 chapters to parts of speech, sentence types, punctuation, and usage. Part II, "Writing the Business Letter," devotes 11 chapters to writing principles, letter mechanics, reports, and several types of letters--including inquiry, and request, order, complaint and adjustment, credit and collection, administrative, sales, and application. Many simple correction exercises are provided along with writing assignments and cases. An appendix includes forms of addresses and salutations.

652 Robertson, Mary, and Perkins, W. E. Practical Correspondence for Colleges. 4th ed. Cincinnati: South-Western, 1974. Index. 210p.
This textbook is intended to guide students in the techniques and psychology of business letters. The inside front cover, regarding the importance of clear business communication, is addressed to the student. The inside back cover has a 25-point checklist of things the student should be able to do after going through the text. The ten units cover such topics as letter format, courtesy, organization, patterns for effectiveness, types of letters, memos and reports, and dictation. At the end of each unit are exercises on grammar and punctuation, study questions based on the unit, rewriting exercises of sentences, and writing assignments of letters based on given situations. There are three appendixes on grammar and punctuation. Appendix C is an answer key.

653 Robertson, W. S., and Siddle, W. D. Technical Writing and Presentation. London: Pergamon, 1966. Index. 118p.
This book is intended for technical people who must write reports, papers, or articles as a part of their work. The 11 chapters include "Planning the Work," "Finding the Right Words," "Illustrations," and "Editing Technical Writing." Chapter 5, "Some Examples Discussed," contains three real reports, which are critiqued. Chapter 11, "Miscellaneous Problems," discusses publishing.

654 Robinson, David M. Writing Reports for Management Decisions. Columbus, Ohio: Merrill, 1969. Index. 407p.
This book focuses on both the principles of writing reports and on the people who write and receive reports--their prejudices, needs, and attitudes. Part I examines the role of the report writer and the report itself. Part II gives the preliminary steps of report preparation (statement of problem, objective, importance of, authorization for) and research for the report. Part III covers sources to use in research, including observation and experimentation. Part IV discusses report organization, writing style, and format. Part V presents the report (timing, interview, presentation to a group) and the nature and needs of decision makers. The 16 chapters include writing assignments, fill-in exercises, and cases.

655 Rogers, Raymond A. How to Report Research and Development Findings to Management. New York: Pilot Industries,

1973.   32p.
Divided into sections and subsections, this booklet focuses on parts
of the report--the title, problem, objective, and recommendations
or conclusions.   Other subjects include meeting the needs of the
reader, achieving clarity, and revising for conciseness.

656   Roland, Charles G.   Good Scientific Writing: An Anthology.
       Chicago: American Medical Association, 1971.

657   Romine, Jack S. ; Hanson, Ladine; and Holdridge, Thelma.
       College Business English.   2nd ed.   Englewood Cliffs,
       N. J. :   Prentice-Hall, 1972.   Index.   386p.
This textbook, which can also be used for self-study, aims to build
proficiency primarily at the sentence level.   The book's 26 "Lessons"
are divided into four parts: "Sentence Construction, " "Grammar, "
"Punctuation, " and "Writing Business Communications. "   The lessons
within each part end with sentence exercises.   The last of the four
parts covers the paragraph, memorandums, and personal business
letters.   The book concludes with eight appendixes, covering such
topics as capitalization, frequently confused words, common business
terms, and the answers to the sentence exercises.

658   Roodman, Zelda, and Roodman, Herman S.   Effective Business
       Communication.   Toronto: McGraw-Hill, 1964.   Index.
       220p.
This textbook of 11 chapters covers such topics as "Claim and Adjustment Letters, " "Credit and Collection Letters, " "The Office Memorandum, " and "The Personal Message in Business. "   Chapter 10,
"Economy in Communication, " discusses cutting costs in letters and
telegrams.   Chapters end in writing assignments.

659   Rose, Lisle A. ; Bennett, Barney B. ; and Heater, Elmer F.
       Engineering Reports.   New York: Harper and Brothers,
       1950.   Index.   341p.
This book of 28 chapters is for "the engineering student and the engineer on the job" who must convey "facts and judgements to fellow
engineers, industrialists, businessmen, and laymen. "   The author
defines effective reports as "products of sound craftsmanship and
contributions to practical action. "   The book opens with an introductory chapter on the roles of reports and engineers within business.
Part I covers information gathering, purpose, and audience.   Part
II treats language effectiveness, mechanics, and illustrations.   Part
III covers business correspondence and summaries and Part IV covers reports.   Part V covers oral reports.   Part VI deals with preparing and marketing articles.   The appendix is a guide to preparing
instructional manuals.

660   Rose, Randolph.   Letters That Make Money.   Lansing, Mich. :
       Business Building, 1915.   63p.
This book is designed to help people in business improve the salesmanship of their letters.   The 12 chapters cover such topics as
attention-getting techniques, effective style, physical appearance,
persuasive techniques, creating desire for a product, effective clos-

ings, the "you attitude, " trite phrases, and friendly tone. Chapter
10 discusses how sales people's calls can be supplemented and re-
inforced by letters.   Chapter 12 discusses the need to personalize
letters--anything that suggests a form letter should be avoided.

661 Rosenblatt, S. Bernard; Cheatham, T. Richard; and Watt, James
     T. Communication in Business. Englewood Cliffs, N. J. :
     Prentice-Hall, 1977. Index. 370p.
This textbook deals with the broad spectrum of business communication--
communication process, speaking, body language, visuals, organizational communication, and the like.   The 15 chapters include "Business Letters" and "Business Reports and Miscellaneous Forms. "
There are discussion questions, problems, writing assignments, and
suggested readings after each chapter.   The eight appendixes cover
favorable, neutral, and unfavorable letters; persuasive letters; looking
for employment; application letters; résumés; and formal report format.

662 Rosenstein, Allen B. ; Rathbone, Robert R. ; Saxneerer, William
     G. Engineering Communications. Prentice-Hall Series
     in Engineering Design. Fundamentals of Engineering De-
     sign. Englewood Cliffs, N. J. : Prentice-Hall, 1964.
     129p.
This book of 11 chapters is designed to illustrate the communication
process--writing, speaking, graphics--for engineers.   The first three
chapters are concerned with the theory of communication and communication systems.   Chapter 4, "The Reader, " covers reader analysis
and the effectiveness of message.   Chapter 5 treats the purposes and
communicative effectiveness of the report.   Chapter 6, "Oral Reporting, " compares oral with written presentation.   Chapters 7-10 are
devoted to graphics--forms, how to sketch, how to use math con-
structions.   Chapter 11 covers methods of illustrating reports.

663 Rosenthal, Irving, and Rudman, Harry W. Business Letter
     Writing Made Simple. rev. ed. Garden City, N. Y. :
     Doubleday, 1968.

664 Ross, Alec. Words for Work: Writing Fundamentals for Tech/
     Voc Students. Boston: Houghton Mifflin, 1970. 129p.
This combination text and workbook has 12 chapters.   Tear-out exercise sheets comprise 33 pages of the book: some are fill-in,
others are sentence correction and paragraph theme exercises.   The
first five chapters review the basic sentence: subjects, verbs, com-
pleters, common problems, and modifiers.   Chapters 6-7 discuss
outlining and logical order, and Chapter 8 is on paragraphing.   Chapter 9, "Stating Your Purpose, " covers introductions and conclusions
as well as determining the purpose.   Chapter 10 treats definition,
explanation, description, narration, the technical report, the business
letter, and the résumé.   Chapter 11, "Visual Forms, " covers outlines as well as illustrations, and Chapter 12, the basics of using
the library.

665 Ross, Alec, and Plant, David. Writing Police Reports: A

Practical Guide. Schiller Park, Ill.: Motorola Tele-
programs, 1977. 100p.
This guide is intended for those who want to be police officers and
dread writing. The nine chapters which have a chatty and informal
style, include "Writing Your Narrative," "Learning to Revise," and
"Writing a Description." Chapter 9 deals with the techniques of taking statements; the last chapter offers words of encouragement for
the reader.

666 Ross, H. John. How to Make a Procedure Manual. 4th ed.
     Miami: Office Research Institute, 1958. 123p.
This book uses the format of a procedures manual, including typewriter font, $8\frac{1}{2}$" x 11" pages, and marginal keys; however, much of
the writing is personal and informal. In Chapter 1, "Definitions and
Purposes," the author does provide a formal definition of the word
"procedure": "A formal instruction which controls the mechanics by
which clerical routines are performed; including equipment, forms
and forms flow, sequence of operations and working conditions."
Chapter 2 covers the design of procedures manuals, and Chapter 3,
writing style, especially clarity and simplicity. Chapters 4-5 discuss the outer covers and bindings for manuals and printing methods.
Chapters 6-7 review internal numbering and types and examples of
content. Chapter 8 presents information on the dissemination of
manuals. An appendix provides procedures sheets for various types
of organizations.

667 Ross, Peter Burton. Basic Technical Writing. New York:
     Crowell, 1974. Index. 349p.
This textbook consists of nine chapters covering the basics of technical writing: reducing raw material into usable information, putting
information into proper order, constructing the three-part summary
and the process summary, and handling such basic items as brief
reports, abstracts, proposals, and long reports. The six chapter-
like appendixes contain information on editing, writing business letters, documenting reports, preparing illustrations, and using abbreviations.

668 Rossen, Harold J. Principles of Specification Writing. New
     York: Reinhold, 1967.

669 Rowland, Dudley H. Handbook of Better Technical Writing.
     Larchmont, N.Y.: Business Reports, 1962. Index.
     239p.
This book concentrates exclusively on grammar, punctuation, and
usage items in 16 chapters. The appendix contains a list of abbreviations of scientific terms approved by the American Institute of
Physics. The illustrations relate primarily to metallurgy.

670 Royds-Irmak, D. E. Beginning Scientific English. Book I.
     London: Nelson, 1975. 146p.
This textbook is for non-native speakers of English. There are 22
units on science topics, such as natural and synthetic rubber, expansion of liquids and gases, and the growth of seeds. Each unit

presents the subject in simple, everyday language and then in more scientific terms. Exercises cover vocabulary, rewriting sentences, answering questions on the text, further discussion, and further activities. Sections of "Revision Exercises" appear at intervals for general review. There are two vocabulary lists: one gives words from each unit, and the other gives the simple English equivalents of terms and phrases in the scientific versions of the units. At the end of the book are notes for teachers who wish to have additional discussion of the questions.

671   Russell, T. H.   Business Correspondence and Forms.   Minneapolis: International Law and Business Institute,   1910.

672   Rusthoi, Daniel.   Prevocational English.   Texts 1 and 2.   Silver Spring, Md.:   Institute of Modern Languages, Text 1:   1970; Text 2:   1971.

673   Rutter, Russell, and Gwiasda, Karl E.   Writing Professional Reports: A Guide for Students.   Dubuque, Iowa: Kendall/ Hunt, 1977.   Index.   99p.
The first three of the seven chapters of this book cover the problem and principles of technical writing as well as definition reports. Chapters 3-7 cover classification and descriptive reports, proposals, and long reports with samples of 13 student reports with critical comments by the authors.   Fourteen exercises follow the chapters -- primarily report exercises.   The book was written as a classroom supplement to W. Paul Jones's Writing Scientific Papers and Reports.

674   Ruxton, Robert.   The Art of Resultful Letter Writing.   2nd ed. Cleveland: Mailbag, 1918.   56p.
The seven chapters in this book were originally published in The Mailbag, a magazine devoted to direct-mail advertising.   The introduction is by Tim Thrift, editor of The Mailbag.   The author is identified as "Chief of Copy Staff of one of the great Direct-by-Mail advertising organizations of America."   (The organization is not identified.)   Chapters are filled with anecdotes of direct-mail advertising experiences.   Topics include physical appearance of letters, making an impression, need for good printing and typing, inquiry letters, sales letters, and follow-up letters.

675   Ryan, Charles William.   Writing: A Practical Guide for Business and Industry.   New York: Wiley, 1974.   Index. Bib.   256p.
This textbook, designed for a technical professional on the job, is a programmed self-study guide.   The 18 chapters are divided into three parts covering writing procedures; larger elements, such as organization, style, and production techniques; effective writing, including grammar, editing, and usage; and basic documents, such as manuals, research reports, specifications, and proposals.

676   Sachs, Harley L.   How to Write the Technical Article and Get It Published.   Washington, D.C.:   Society for Technical

Communication, 1976.   64p.
This guide is intended for those who know how to write effectively
but need advice on publication.   The 14 chapters are divided into
three parts: "Gathering Information, " "Writing the Article, " and
"Selling What You Write. "   Chapter topics include library research,
interviewing, titles, leads and endings, rewriting, preparing and
selling the manuscript, and writing while traveling.   Chapter 14,
"The Business of Writing, " covers work space, record keeping, and
tax tips.

677   Sales Correspondence.   Chicago: Shaw, 1916.

678   Sales Promotion by Mail: How to Sell and How to Advertise.
New York: Putnam's, 1916.   Index.   359p.
According to the foreword, this collection of ten chapter-length articles, each written by an "editor, " "deals with every phase of getting business through the mails. "   The book begins with a chapter on
compiling a mailing list.   The next five chapters deal with specific
kinds of letters: form, follow-up, collection, response to inquiry,
and letter enclosures (flyers, promotional pieces, etc.).   The next
two chapters review organizing advertising and promotion departments.   Chapter 9 covers planning and editing a house organ and
building an export business by mail.

679   Salzberg, Richard, ed.   Professional Advancement for Engineers
Through Communication Techniques.   Los Angeles: Par-
sons, 1968.

680   Santmyers, Selby S.   Practical Report Writing.   Scranton, Pa.:
Laural, 1950.   Index.   Bib.   120p.
This book aims to help those "persons--engineers, salesmen, accountants, production supervisors, public officials, or officers of
labor unions, fraternal organizations, or service clubs--write reports. "   Divided into eight chapter-length sections, the book numbers
every topic within those sections--139 in all.   The first section discusses the "Importance of Writing Reports, " their qualities, and who
writes them.   The remaining sections of the book cover organization,
the parts of a report, the introduction, the body, paragraphing, and
conclusions.   The book includes writing exercises in each section.
The last section of the book provides sample report pages, format
advice, and a brief list of recommended books.

681   Sarma, G. V. L. N.   English for Engineering Students.   New
York: Asia Publishing House, 1964.   Index.   235p.
This textbook of 11 chapters is an ESL guide for engineers and engineering students in India.   The first five chapters deal with vocabulary, spelling, punctuation, grammar, and usage.   Chapter 6 reviews
choice of words, word order, sentence structure, and paragraph
writing.   Chapters 7-10 cover various types of writing: letters, abstracts, essays, and reports.   Chapter 11 is a guide for learning
English through paraphrasing.   Although the chapters on types of
writing provide rhetorical strategies, their focus is on mechanics
and structure.   The exercises consist of sample passages that stu-

dents are asked to correct. The chapters are followed by two appendixes: a pronunciation guide and a reading list of English works,
biography and fiction mostly.

682 Saunders, Alta Gwin. Effective Business English. 3rd ed.
         New York: Macmillan, 1949. Index. Bib. 871p.
This textbook of 22 chapters is intended for both students in colleges
and employees on the job. Professor Francis Weeks wrote "the chapters on goodwill and business reports, and the 'Handbook of English'
(Appendix I). " Professor Hugh W. Sargent prepared the exercises
and letter problems. The first six chapters of the text are concerned
with the "Cardinal Qualities of Business Letters" and the planning of
effective letters. Chapters 7-21 deal with different kinds of letters:
sales, job-application, credit, collection, claim, goodwill, etc. Chapter 22 treats report writing. There are five appendixes on such
matters as "letter cost, " letter appearance, and "Correct Letter
Salutations. " In addition to the writing assignments and letter problems, several types of letters are critiqued.

683 Saunders, Alta Gwin, and Creek, Herbert LeSourd, eds. The
         Literature of Business. rev. ed. New York: Harper
         and Brothers, 1923. Bib. 554p.
This collection is aimed at students in business writing at colleges
and universities or "correspondents" being trained in companies.
The theory of the collection is that a business letter is an extension
of personality, and, further, an extension of the personality of a
business. Part I, "The Profession of Business, " contains readings
on education, ethics, psychology, biography, and success. These
readings emphasize the cultivation of a successful, cultured business
personality in addition to training in business theory. Part II, "Business Writing and Related Principles of Business, " contains readings
on the general functions of letters, claims and adjustments, credits
and collections, applications and "positions, " advertising and salesmanship, and other types of business writing. Authors represented
in the collection include Samuel Johnson, John Ruskin, Theodore
Roosevelt, John D. Rockefeller, Henry Ford, and John Opdycke.

684 Sawyer, Thomas M. , ed. Technical and Professional Commu-
         nication. Ann Arbor, Mich.: Professional Communica-
         tion, 1977. 204p.
The description on the book's cover, "Teaching in the Two-Year
College, Four-Year College, and Professional School, " accurately
describes the purpose of this anthology. In his introduction, the
editor suggests that this collection aims to help the beginning instructor in technical communication. He asserts that English departments have for the most part failed to assume this gigantic task.
The scope of the 20 articles is broad, covering not only standard
pedagogical approaches but also teacher preparation, publishing, and
internships. One article provides a bibliography for beginning teachers and another discusses the tradition of scientific and technical
writing. A brief description of each contributor is included.

685 Sawyer, Thomas S. Specification and Engineering Writer's

Manual. Chicago: Nelson-Hall, 1960. Index. Bib.
231p.
The 19 chapters of this book are fairly short, each ending in a summary and "work program" (study/thought-provoking questions). The
book is divided into five parts. Part I contains writing techniques,
usage, and a discussion of specifications and other engineering documents and their origins. Part II presents the process for preparing
specifications: research, writing, editing, and distribution. Part
III covers the legal aspects of specifications. Part IV discusses
equipment and office procedures then closes the book with a pep talk.
Part V provides 70 pages of examples of construction and installation
specifications, proposals, reports, operation manuals, and a table of
standard abbreviations.

686 Scholl, C. Phraseological Dictionary of Commercial Corres-
     pondence in the English and French Languages. London:
     Hachette, 1911.

687 Scholl, C. Phraseological Dictionary of Commercial Corres-
     pondence in the English and Spanish Languages. London:
     Hachette, 1911.

688 Schultz, H., and Webster, R. G. Technical Report Writing:
     A Manual and Source Book. New York: McKay, 1962.
     Index. 359p.
This textbook is designed for upper-level students in four-year
schools. The authors use examples from such fields as physics,
chemistry and chemical engineering, geology, civil and electrical
engineering, industrial management, government, and "insurance"
engineering. About one-third of the book is devoted to four parts
covering "Aims," "Appearance," "Styling," and the "Language" of
formal and informal reports. The second, larger, section of the
book deals with "uniquely practical source material," containing case
studies or problems for students to develop into full-fledged reports.
There are 50 problems from various fields; of these, two sample
reports "are worked out in full to guide in handling the others." A
five-page appendix of standard abbreviations of "Scientific and Engineering Terms" is included.

689 Schutte, W. M., and Steinberg, E. R. Communication in Busi-
     ness and Industry. 1960; rpt. Huntington, N.Y.: Krieg-
     er, 1974. 386p.
This book, a reprint of an edition published by Holt, Rinehart and
Winston in 1960, is an informal, humorous treatment of business
writing and speaking. Topics include writing letters, memos, and
reports and using illustrations. The book uses a workbook, self-
teaching approach. Each chapter states a problem, gives good and
bad examples, provides entertaining examples of the principles, and
asks questions for the user to answer.

690 Scott, John Hubert. Engineering English. New York: Wiley,
     1928. Index. Bib. 321p.
According to its author, "the purpose of this book is to aid the stu-

dent of engineering in his efforts to develop ... excellence of expression. " The author, in his introduction, also differentiates English from "Engineering English. " The 25 chapters are divided into
two parts. Part I treats some general principles, products, and
attention to detail in Engineering English: the "process paper, "
"direction paper, " "apparatus paper, " letter writing, report writing,
and mechanics. Part II covers other forms of Engineering English,
written and spoken: "the definition paper, " "the new-invention talk, "
"the investigative paper, " "the argumentative paper, " "the engineering-
project talk, " "the thought paper" (abstract ideas), and specification
writing. The author provides no exercises or other pedagogical devices. Examples are primarily from civil engineering.

691 Seybold, Geneva. Communicating with Employees About Mer-
          gers. New York: National Industrial Conference Board,
          1968. 59p.
This report, divided into three parts, discusses the timing and selection of various media (e. g. , news releases, personal letters,
memos, and bulletins) when dealing with mergers. It describes
specifically the timing and the selecting of media for nine companies
that merged with larger organizations. Part I describes the objectives of these companies, Part II presents the case histories, and
Part III is a summary for the executive. The report concludes with
an index of tables.

692 Seybold, Geneva. Personnel Procedure Manuals. Personnel
          Policy Study, No. 180. New York: National Industrial
          Conference Board, 1961. Index. 123p.
This research report, which is a part of a series of personnel policy
studies, examines the content, format, preparation, and distribution
of personnel procedure manuals. The first two chapters discuss the
contents of manuals; the next two chapters cover organization and
format. Chapter 5, "Preparation of Statements, " treats the writing
and review of statements in the manual. Chapter 6 discusses the
distribution of manuals, and Chapter 7, their revision. Chapter 8
treats how to train supervisors to use manuals and provides samples
from various manuals. The appendix contains a composite subject
index derived from a vast collection of personnel manuals. The
book's table of contents appear at the end.

693 Shaaber, M. A. The Art of Writing Business Letters. Boston:
          Houghton Mifflin, 1931. 431p.
This textbook of 12 chapters covers such letters as sales, collection,
adjustment, and rejection of claims. Chapters 1-3 treat general
principles of clear writing and building goodwill through letters.
Chapters 4-7 cover the layout of long and short letters. Each chapter is followed by a section containing questions, cases on business
problems, and rewrite exercises.

694 Shaw, James, ed. Teaching Technical Writing and Editing--
          In-House Programs That Work. Washington, D. C. : So-
          ciety for Technical Communication, 1976.

695 Shearing, Henry Arthur, and Christian, B. C. Reports and
        How to Write Them. London: Allen and Unwin, 1965.
        Index. Bib. 141p.
This book is designed for managers and technicians who write reports
for their organizations. Chapters 1-3 deal with the purpose and function of reports; Chapters 4-5 treat structure and writing effective
reports, as well as clarity, appropriateness, and readability. Chapter 6 deals with grammar, usage, and common stylistic pitfalls in
report writing.

696 Sheff, Donald A. Business English. New York: Ronald, 1964.
        Index. 292p.
This book with tear-out worksheets contains 11 lessons on grammar
and punctuation principles. The author states that the chief purpose
of the book is to increase the "grammatical know-how" of students.
After Lesson 11, there is a list of common abbreviations. Each
lesson has abundant exercises giving practice with the principles.
Examples are not from a business context.

697 Shepherd, Ray. This Business of Writing. Chicago: Science
        Research Associates, 1980. 224p.
This business-English workbook is designed to sharpen students' basic
writing skills. The six parts are divided into modules on such topics as sentence structure, grammar and usage, punctuation, vocabulary, and business letters. The workbook begins with a 140-item
"Diagnostic Grammar and Usage Test." Students fill in answers to
questions for each module and check themselves in answer keys.
Part VI is a reference handbook of grammar and usage.

698 Sheppard, Mona. Plain Letters: The Secret of Successful
        Business Writing. New York: Simon and Schuster, 1960.
        Index. Bib. 305p.
This book is written for people in business and government who wish
to write clear, or "plain," letters. After a brief opening chapter,
the book is divided into four parts. Part I, "How to Write a Plain
Letter," discusses the "4-S Formula" (Shortness, Simplicity, Strength,
and Sincerity) and dictating. Part II, "The Practical Approach to
Grammar and Punctuation," treats 40 usage problems and the comma.
Part III, "Production Aids," covers physical layout and the use of
form letters. Part VI, "Examples," provides 60 sample letters.
At the end is a four-part appendix: a self-inventory called "Test
Your L. Q. (letter-writing quotient), a "watch list" of outdated business-
letter terms, a "forms of address" section, and a brief bibliography.
Each of the 11 chapters concludes with a summary.

699 Sherman, Theodore A., and Johnson, Simon S. Modern Tech-
        nical Writing. 3rd ed. Englewood Cliffs, N. J.: Prentice-
        Hall, 1975. Index. Bib. 480p.
This textbook, aimed at students in engineering and science, has 16
chapters divided into four parts. Part I, "Technical Writing in General," covers diction, sentence structure, organization, manuscript
style, special formats, and graphics. Part II treats "Reports, Proposals, and Oral Presentation." Part III, "Business Correspondence,"

covers the basic types of letters, reader benefits, and format. Part
IV, "Handbook to Fundamentals," is a guide to grammar, mechanics,
and punctuation. In addition to a glossary of usage and a glossary
of grammatical terms, there are two appendixes: one is a list of
technical abbreviations and the other a bibliography for further reading. Each chapter has writing exercises for the student, ranging
from sentence exercises to lengthy writing assignments. All of the
numerous sample reports and letters are shown in facsimile, and
all were produced by professional people who also had to write.

700   Shidle, Norman G.   The Art of Successful Communication:
          Business and Personal Achievement Through Written
          Communication.   New York: McGraw-Hill, 1965.   Index.
          267p.
This book is designed for people in business who must write letters,
reports, articles, speeches, minutes of meetings, and memoranda.
It covers especially persuasive writing, good- and bad-news messages, and requests for information.   The book begins with a light
treatment of the theory of and barriers to communication.   It follows
with a general discussion of how writing improves communication and
how writers can focus on their readers.   Of the 24 chapters, eight
are devoted to specific kinds of business communications, such as
memos and minutes of meetings. Also discussed are writing speeches
and articles for technical and business magazines.   The final chapters cover listening, reading, and the satisfaction gained from effective communications.

701   Shryer, W. A.   Collecting By Letter.   2 vols.   Detroit: Busi-
          ness Service, 1913.

702   Shulman, Joel J.   Treasurer's and Controller's Letter Book.
          Englewood Cliffs, N. J. :   Prentice-Hall, 1974.

703   Shuman, John T.   English for Vocational and Technical Schools.
          2nd ed.   New York: Ronald, 1954.

704   Shurter, Robert L.   Effective Letters in Business.   2nd ed.
          New York: McGraw-Hill, 1954.   Index.   250p.
This textbook covers the fundamentals of major types of business
letters: claim and adjustment, credit, collection, sales, and application.   Chapter 1 discusses "What Is an Effective Letter?" and
covers such principles as tone, the "you attitude," and personality.
The 12 chapters also review business jargon, memos, letter forms,
and clarity.   The chapters are filled with short examples of various
types of letters.   Each chapter ends with exercises, which include
correction of sample letters, punctuation of business sentences, and
writing assignments for specific situations.

705   Shurter, Robert L.   Principles of Business Communication.
          New York: McGraw-Hill, 1957.

706   Shurter, Robert L.   Written Communication in Business.   3rd
          ed.   New York: McGraw-Hill, 1971.   Index.   Bib.   623p.

This textbook is "designed to present a comprehensive treatment of
the major principles of business communication, an analysis of the
most widely used forms of business writing--the letter, the report,
and the memorandum--and a discussion of the associated skills of
dictation and reading. " Aimed at both college students and business
people, this book has 22 chapters, divided into five parts: "Principles of Business Communication, " "Techniques of Business Letters, "
"Specific Types of Business Letters, " "The Report and Memorandum, "
and "Self-Development on the Job. " The chapters contain numerous
exercises and discussion questions; an appendix contains cases. Following the chapters is a 95-page handbook section on usage, grammar,
punctuation, and mechanics.

707 Shurter, Robert L.; Williamson, J. Peter; and Broehl, Wayne
         G. , Jr.   Business Research and Report Writing.   New
         York: McGraw-Hill, 1965. Index. Bib. 204p.
This book of ten chapters is designed as a guide for students and
those on the job.   Chapter 1, "The Purpose of a Report, " is devoted
to the rhetorical analysis of audience.   The next three chapters,
"Clarity and Coherence, " "Writing Correctly, " and "Styled to the
Reader's Taste, " cover general writing principles.   Chapter 5, "Research Methods, " lists sources and logic in the correct use of information.   Chapter 6, "Statistics in Business Research, " provides
a quick review of the presentation of numerical information and data.
Chapters 7-9 cover organization and writing the major report elements, including beginnings and endings.   Chapter 10, "Graphic
Presentation, " is taken from the AT&T booklet written by Kenneth
W. Haemer.

708   Sieff, M.   Manual of Russian Commercial Correspondence.
         New York: Dutton, 1916.

709   Sigband, Norman B.   Communication for Management and Busi-
         ness.   2nd ed.   Glenview, Ill.: Scott, Foresman, 1976.
         Index.   657p.
This textbook is designed for a one-semester course in business
communication (letters and reports) or a two-semester course in
business communication (theory, research, principles, letters, and
reports).   The major divisions of the text deal with the theory and
process of business communication and varieties of business letters
and reports.   A final section is composed of readings in organizational communication.   The 20 chapters are fully illustrated with
examples from business, and each chapter has detailed exercises
for discussion and writing assignments.   There are also chapters
on career planning and résumés.   There are two appendixes: the
first is a brief grammar and punctuation handbook, and the second
covers the typing and format of business letters.

710   Sklare, Arnold B.   Creative Report Writing.   New York:
         McGraw-Hill, 1964.   Index. Bib. 428p.
This book is designed for people in business, professions, government, or colleges who want to improve their ability to communicate
in their professions.   In his definition of report writing, the author

equates creativeness with inventiveness.   The 26 chapters include
a comparison of technical writing to belles lettres and other forms
of writing, such as journalism.   In addition, topics include meaning
and language, the writing process, graphic devices, and special
kinds of reports (e. g. , accounting, medical, and legal).   There are
two appendixes: a bibliography of report writing and a section on
research techniques.

711   Sklare, Arnold B.   The Technician Writes: A Guide to Basic
              Technical Writing.   San Francisco: Boyd and Fraser,
              1971.   314p.
This textbook of 17 chapters is aimed at the student in a technical
school.   The emphasis is on "the basic and the practical. "   Part I
devotes ten chapters to the technical report, including library research; Part II has four chapters on letters, memos, articles, and
abstracts; Part III includes four chapters on functional English, sentences, paragraphs, and grammar; Part IV is a two-part handbook
section on usage and grammatical terms.   Although there are no
exercises or writing assignments, several readings are included.

712   Slattery, James.   Business Letter Writing.   A Tutortext Pre-
              pared Under the Direction of Educational Science Division,
              U. S. Industries, Inc.   Garden City, N. Y. :   Doubleday,
              1965.   Index.   246p.
This book is designed to be used as a self-instruction guide for anyone interested in business communication.   It could also be used in
company in-house programs.   There are seven chapters covering
effective sentences, effective word choice, the purpose of business
letters, tone, sales messages, and organization.   Chapter 7 is a
collection of business letters that the reader is to revise.   The appendix explains a "Business Communication Efficiency Index, " which
is meant to help readers evaluate their own letters and also contains
a section on correct forms of address.   The book is not designed to
be read from beginning to end.   Rather, the material is divided into
numbered sections and readers find instructions in each section regarding what material to read next.   For example, on page one there
is a selection of definitions of a good business letter.   Readers select a definition and then check themselves by turning to the section
indicated, where they are told whether or not they are correct and
why.   They are then sent on to another section.

713   Slocum, Keith D.   Business English: A Worktext with Pro-
              grammed Reinforcement.   2nd ed.   Indianapolis: Bobbs-
              Merrill, 1980.

714   Smart, Walter K.   How to Write Business Letters.   Chicago:
              Shaw, 1916.

715   Smart, Walter Kay; McKelvey, Louis W. ; and Gerfen, Richard
              C.   Business Letters.   4th ed.   New York: Harper and
              Brothers, 1957.   Index.   603p.
This textbook of 15 chapters, designed for business students, begins
with chapters on the approach to, the writing of, and the forms for

business letters. Eleven chapters then deal with types of business
letters--including order and acknowledgment, claim and adjustment,
credit, collection, sales, goodwill, form and guide, and application.
Chapter 14 covers reports. The last chapter is a handbook of business English with exercises. Cases and writing assignments are
included.

716 Smith, Arthur M. Writing Readable Patents, Including "The
        Art of Communicating Complex Patent Matters" by Harry
        C. Hart. New York: Practising Law Institute, 1958.
        64p.
This work is actually two monographs bound together. The first and
largest, by Arthur M. Smith, is addressed to patent attorneys. This
monograph encourages attorneys to "develop better writing skills"
and "apply the principles and techniques of good writing ... to the
writing of patents." Throughout the monograph, the author uses examples that come from successfully written patents. In addition to
writing style, the book covers strategy for composing specific parts
of patent applications: "summarizing the Invention" and "naming the
parts." Throughout, the author advises avoiding technical language
whenever possible. The second monograph, by Harry C. Hart, covers the theory for discovering the uniqueness of a particular patent
and then communicating that uniqueness with unity and coherence.

717 Smith, Carrie J., and Mayne, D. D. Modern Business English.
        Chicago: Powers and Lyons, 1906. 256p.
This textbook, designed for the first year of high school and for
business colleges, has 97 chapters. Chapters on letter writing alternate with chapters on grammar and punctuation. Topics include
"Sentence Sense," "Parts of Speech--Noun," "Importance of the Letter," "Dunning Letters," "Circular Letters," and "Advertisements."
Four appendixes cover abbreviations and postal information. Each
chapter ends in sentence exercises or writing assignments.

718 Smith, Charles B. Practical Word Choice in Business Writing.
        4th ed. Dubuque, Iowa: Kendall/Hunt, 1978.

719 Smith, Leila R. English for Careers. New York: Wiley,
        1977.

720 Smith, Patrick D., and Jones, Robert C. Police English: A
        Manual of Grammar, Punctuation and Spelling for Police
        Officers. Springfield, Ill.: Thomas, 1969. 77p.
This book of three chapters covers grammar, punctuation, and spelling, with examples drawn from police work. Chapter 1, "Grammar
of the Sentence," covers basic sentence patterns, transformations,
and rules for combining and subordination. Chapter 2, "The Mechanics and Punctuation of a Report," covers only punctuation. Chapter
3 is "Spelling."

721 Smith, Randi Sigmund. Written Communication for Data Pro-
        cessing. New York: Van Nostrand Reinhold, 1976. In-
        dex. 199p.

This book covers the general principles of written communication, but it applies those principles specifically to systems development, operations, modification, maintenance, and management documents in data processing. The chapters cover such topics as "making valid assumptions about the reader," "the think through," brevity, word selection, plain speech, and salesmanship. The chapters also cover six kinds of "memos" used in the field: presentation memos, status reports, information memos, management reports, review memos, and inquiry memos. Throughout the book, the author wages war on computer jargon and its mutations.

722 Smith, Richard W. Technical Writing. New York: Barnes and Noble, 1963.

723 Smith, Terry C. How to Write Better and Faster. New York: Crowell, 1965. Index. Bib. 220p.
According to its foreword, "this book is for everyone who writes -- businessmen (particularly the junior executive), government employ- ees, engineers, secretaries, and even housewives...." The ten chapters cover such topics as prewriting and effective style, editing techniques, letters and memos, reports and proposals, advertising, publicity, public relations, and speech writing. Chapters 8-9 cover mechanics and illustrations. Chapter 10 provides an overview of printing and duplication processes. A glossary of usage and a list of suggested readings are included.

724 Sollers, George W. The Policeman's Guide to Report Writing. N.p., 1916.

725 Souther, James W. Exposition for Science and Technical Stu- dents. New York: Sloane, 1950.

726 Souther, James W. A Guide to Technical Reporting. rev. ed. Seattle: University of Washington Press, 1954. 109p.
This textbook of six chapters applies the engineering method to the report-writing process. Chapters cover "Letters and Correspon- dence," report forms and types, references used in compiling re- ports, organization, layout, style of reports, and abstracts. The three appendixes cover speaking and provide sample letters and re- ports. Exercises and study questions follow.

727 Souther, James W., and White, Myron L. Technical Report Writing. 2nd ed. New York: Wiley, 1977. Index. Bib. 93p.
According to the authors, this book is a "blending of theory and philosophy ... with practical dimensions of writing." The early chapters treat the principles of technical and scientific writing and the basic writing process. Later chapters discuss communication problems, forms of rhetorical presentation, report design, and ap- plications of writing and revision. The book ends with an essay on "the point of diminishing returns in writing."

728 Sparrow, W. Keats, and Cunningham, Donald H. The Practical

Craft: Readings for Business and Technical Writers.
Boston: Houghton Mifflin, 1978.   Index.   306p.
This collection of 28 articles is divided into five divisions: "What
Is Business and Technical Writing and Why Study Such Writing?";
"What Style Is Appropriate for Business and Technical Writing?";
"What Are Some Important Writing Strategies?"; "What Are Some
Important Types of Letters and Reports?"; and "What Are the Im-
portant Formal Elements of Reports?"   Each division has a brief
introduction giving an overview of the topic and includes an annotated
bibliography of further readings.   Some of the authors included are
Stuart Chase on gobbledygook, Christian K. Arnold on writing abstracts, Lois DeBakey on the persuasive proposal, and Rudolf Flesch
on readability.   Discussion questions follow each article.

729   Sperling, Jo Ann.   Job Descriptions in Marketing and Manage-
            ment.   AMA Research Study 94.   New York: American
            Management Association, 1969.   Index.   223p.
Written to help executives prepare and evaluate job descriptions, this
report is divided in four sections.   The first provides an overview
of the findings of an AMA survey on job descriptions.   The second
discusses the purpose, use, and importance of job descriptions.   The
third section gives advice about writing the job description: techniques, who should write it, and other related factors.   The fourth
section discusses job-description format and the content of various
sections.   The remaining 200 pages are a collection of actual and
composite job descriptions divided into two groups.   The first group
illustrates job descriptions in marketing from two companies: General Mills, Inc., and a manufacturer of electrical products.   The
second group shows 29 job descriptions of positions taken from selected companies.   Company names and titles of job descriptions are
indexed.

730   Squires, Harry A.   Guide to Police Report Writing.   Spring-
            field, Ill.:   Thomas, 1964.   84p.
This spiral-bound book covers in four chapters what the author calls
the "nuts and bolts" of police reports.   Chapter 1, "Introduction,"
defines and explains police reports and gives stylistic principles
(clarity, conciseness, etc.).   Chapter 2, "Steps in Report Writing,"
covers note taking.   Chapter 3, "Writing Techniques," discusses
simple word choice in three pages.   Chapter 4, "Writing Rules,"
gives definitions of grammar terms, spelling rules, punctuation, and
usage.   Chapters 1, 2, and 4 have exercises.   There are five appendixes: recommended abbreviations, definitions of criminal terms,
a reading list, canons of police ethics, and a checklist for reports
and investigations.

731   Stephenson, Howard.   What an Executive Should Know About
            Effective Report Writing.   Chicago: Dartnell, 1973.

732   Stevens, J. D.   Writing Better Titles and Abstracts.   Pullman:
            Washington State University Press, 1961.

733   Stevenson, Brenton W.; Spicer, John R.; and Ames, Edward C.

English in Business and Engineering. New York: Prentice-
Hall, 1936. Index. 365p.
This textbook of ten chapters covers the psychology of communication,
the business letter, the research article, oral and written reports,
and press releases. The book provides numerous examples and illustrations of reports and a discussion of logical fallacies and persuasive argument. The final three chapters cover grammar, spelling, and punctuation. No exercises or other pedagogical material
are included.

734 Stevenson, J. Principles and Practice of Commercial Correspondence. New York: Pitman, 1958.

735 Stewart, John L. Exposition for Science and Technical Stu-
dents. New York: Sloane, 1950. Index. 258p.
This textbook concentrates on report writing. The nine chapters
cover such topics as "Language and Scientific Exposition," "The
Paragraph," "Reports: Their Forms and Uses," and "Reports:
Their Formats." Each chapter ends in writing exercises.

736 Stewart, Marie M., et al. Business English and Communication. 4th ed. New York: McGraw-Hill, 1972. Index.
Bib. 542p.
This textbook aims to teach high school students "the ability to use
[the English language] effectively for business purposes." The 62
units in three sections provide instruction in writing, language, reading, listening, and speaking. Section I, "Foundations of Effective
Communication," cover word choice, vocabulary, listening, reading,
grammar, and punctuation. Section II, "Effective Written Communication," covers communication psychology, writing style, effective
letter strategy, types of letters, memos, reports, and minor forms
of communication (e.g., telegrams). Section III, "Effective Oral
Communication," treats speech and vocal factors as well as various
types of oral presentations. The appendix provides a spelling list,
state abbreviations, forms of addresses, and a list of standard reference books. Included are exercises, writing assignments, discussion questions, and cases.

737 Stockwell, Richard E. The Stockwell Guide for Technical and
Vocational Writing. Menlo Park, Calif.: Cummings,
1972. 296p.
This textbook is designed for vocational students. The 12 chapters
provide a basic approach to on-the-job writing. In addition to letters
and reports, the chapters cover work orders, maintenance records,
and employment application forms. The appendix includes information on basic abbreviations of technical terms, a brief discussion of
writing style that slants meaning, and a table for handwriting instructions.

738 Stone, Robert. Successful Direct Marketing Methods. Chicago:
Crain, 1974.

739 Stowe, A. Peter [Stoll, Albert, Jr., pseud.]. Making the Letter

Pay.   Detroit, Mich.:   Business Man's Publishing, 1913.
106p.

The author observes in Chapter 4 that "business style does not wholly conform to literary style.   Smooth-flowing, euphonious, well-rounded words, set in long sentences, usually give a tedious tone to the business letter."   Chapter 1 outlines the purpose of a business letter;   Chapters 2 and 3 cover elements of letter form;   Chapter 4, style;   and Chapters 5 and 6, the form letter: its style and distribution.   Chapters 7-13 discuss office procedures, filing, follow-up letters, and classification of correspondence.

740   Stratton, Charles R., ed.   Teaching Technical Writing: Cassette Grading.   Morehead, Ky.:   Association of Teachers of Technical Writing, 1979.   Bib.   45p.

This collection of seven articles is part of a series of guides for teachers of technical writing.   The articles, all published from 1970 to 1976, are followed by postscripts by the authors, giving an update on the techniques being discussed.   Selections include "Cassettes in the Classroom," by Enno Klammer;   "The Use of the Tape Recorder in Grading," by John S. Harris;   and "Technological Gift-Horse: Some Reflections on the Teeth of Cassette-Marking," by Russell A. Hunt.

741   Strong, Charles W., and Eidson, Donald.   A Technical Writer's Handbook.   New York: Holt, Rinehart and Winston, 1971.   Index.   368p.

This handbook is geared to "all the people who must do technical writing--including professional technical writers, journalists, scientists, engineers, administrators, technicians of all kinds, and students of agriculture ... physics."   The 15 chapters include discussions of technical style, proposals, "Format for Government Publications," "Evaluating Data: Statistics," "Technical Editing," and "Technical Articles."   The "Short Guide to References" cites information on references in various scientific and technical specialties. The 128-page index serves as a handbook of usage and glossary as well as an index to topics covered in the book.

742   Strong, Earl P.   Writing Business Letters.   New York: American Book, 1950.   Index.   329p.

This text-workbook with tear-out pages is divided into three parts and an Introduction.   The divisions within the parts are unnumbered. The Introduction discusses planning and the basic principles of business letter writing.   Part I, "Mechanics of the Business Letter," includes stationery, headings, addresses, and envelopes.   Part II, "Fundamentals of Business Letter Writing," treats choice of words, grammar, punctuation, openings, and closings.   Part III, "Applications of Principles of Business Letter Writing," covers order letters, sales letters, etc.   There are numerous fill-in and letter-writing exercises at the end of each section.

743   Strong, Earl P., and Weaver, Robert G.   Writing for Business and Industry: Reports, Letters, Minutes of Meetings, Memos, and Dictation.   Series in Administrative Management.   Boston: Allyn and Bacon, 1962.   Index.   456p.

Despite its title, the major emphasis of this textbook is on letter
writing.   Only Chapter 11 covers reports and other forms of business writing.   In the first of three parts, the book briefly discusses
principles of communication, format advice, and examples of their
application.   Part II treats specific business letter problems and
various kinds of letters in detail--sales, order, credit, job application, and collection.   Part III serves as a reference manual on the
fundamentals of English applied to business writing.   Most chapters
give detailed guidance in sections titled "To Aid You."   All chapters
provide numerous exercises, writing assignments, and cases.

744 Swales, John.   Writing Scientific English.   London: Nelson,
          1971.

745 Swindle, Robert E.   The Business Communicator.   Englewood
          Cliffs, N. J.:   Prentice-Hall, 1980.   Index.   436p.
The 28 chapters of this textbook are divided into seven parts.   Parts
I and II deal with the introduction to the communication process along
with effective style.   Part III treats memos, letters, and telegrams.
Part IV discusses the "Psychological Techniques" used in delivering
good and bad news, and in persuasion.   Part V deals with nonwritten
communication, e. g. , oral, graphics, and nonverbal.   Part VI covers
reports, and Part VII, the job search.   Exercises include class discussion and writing assignments.   Checklists for various forms of
business communication are included.

746 Sykes, J. B. , ed.   Technical Translator's Manual.   London:
          Aslib, 1971.   Index.   Bib.   173p.
The ten chapters of this book were compiled by members of the
Committee of the Aslib Technical Translation Group.   The material
deals primarily with the translation of written work into English,
from the standpoint of a British translator.   The chapters cover
such topics as "The Training of Translators," "The Staff Translator,"
"The Relations Between Languages," and "The Translator's Tools."
Chapter 9 provides a selective list of indexes of existing translations
and of translators.   Chapter 10 is a review of the literature on technical translation.

747 Sypherd, W. O. ; Fountain, Alvin M. ; and Gibbens, V. E.
          Manual of Technical Writing.   Chicago: Scott, Fores-
          man, 1957.   Index.   Bib.   560p.
Designed for both professionals and technical students, this book is
a revision of The Engineer's Manual of English.   The book is divided
into eight chapters, with numerous exercises following each one.
Chapter 1, "Technical Communication," provides the background and
establishes the need for clear technical style.   Chapter 2 reviews
"Correspondence"; Chapter 3, "The Library Research Paper," is
designed for the student.   Chapters 4-6 cover forms of writing on
the job: "Reports," "Writing for Technical Journals," and "Writing
for Company Publications."   Chapter 7 discusses "Illustrative Aids,"
and Chapter 8 is "Speech for the Technical Man."   Following the
chapters is a 72-page reference manual of grammar, usage, and
mechanics.

748 Tallent, Norman. Psychological Report Writing. Englewood
        Cliffs, N. J. : Prentice-Hall, 1976. Index. Bib. 262p.
This book focuses on the relationship of writing to the interpretation
of data: "The theme runs throughout the book that the psychologist,
rather than presenting 'results' of his tests, interacts with his data
and generates conclusions that might be useful in meeting perceived
needs of the client in his personal uniqueness and the uniqueness of
his situation. The generation of such situational conclusions is interpretation. " Part I, in five chapters, stresses the importance of
writing. Chapter 1, "The Psychological Report: Purpose and Content, " examines the role of the psychologist in the report-writing
process and with the health-care team. Chapter 2, "Pitfalls in Reporting, " treats such factors as misuse of language, poor emphasis,
and faulty interpretation. Chapter 3, "Responsibility and Effectiveness, " covers the point of view appropriate to psychological reports.
Chapters 4-5 cover the content and organization of effective reports.
In Part II, Chapters 6-8 include samples of reports and comments.
Following the eight chapters is a list of "References" for report
writing in psychology.

749 Thomas, J. D. Composition for Technical Students. 3rd ed.
        New York: Scribner's, 1965. Index. 461p.
This textbook has 11 chapters divided into four parts. Part I is a
single chapter titled "Good English in Technical Style. " Part II is
composed of four chapters that actually serve as a "Handbook of
Fundamentals" (grammar, usage, punctuation, mechanics). Part III,
"Types of Discourse in Technical Composition, " covers exposition,
description, narration, and argumentation. Part IV, "Forms of
Technical Communication, " covers letter writing, library research,
the technical report, and the technical speech. Two appendixes follow the chapters. Appendix A provides exercises for the 11 chapters, and Appendix B contains readings from the past and modern
times on technical subjects. Following the index is a four-page alphabetical guide to the Handbook (Part II). All chapters provide exercises and discussion questions. In his discussion of style, the
author vigorously defends the use of the passive voice.

750 Thompson, Karl Owen. Technical Exposition. New York:
        Harper and Brothers, 1922. Index. 231p.
This text, based on a course at Case School of Applied Science, is
designed "to cover the more practical of the instruction in English
which follow the ground work in composition and rhetoric and is designed for a single semester's course. " The 15 chapters begin with
a definition of the expository method in its context of technical writing (although that term is not used). The three following chapters
cover the use of words, including their sources and the vocabulary
of technology. Chapter 5 focuses on the sentence, and Chapter 6
covers punctuation, abbreviations, and symbols. Chapter 7 treats
the overall organization of written material, and Chapter 8 covers
oral presentation. The next six chapters discuss types of exposition: the business letter, journalistic articles, advertising copy,
technical reports, and specifications and contracts. The final chapter, "Accessories to Exposition, " presents the use of illustrations
and sources for reference.

751 Thornley, G. C. Scientific English Practice. rev. ed. New
      York: Longman, 1975.

752 Tichy, H. J. Effective Writing for Engineers, Managers, Sci-
      entists. New York: Wiley, 1966. Index. 337p.
The author aims here to present the philosophy of good style and to
provide readers with writing techniques they can use with imagination
and originality. The 16 chapters move from an overview of the book's
design and the four steps of writing (plan-write-cool-revise) to specific problems in organization, outlining, brevity, grammar, word
choice, and development. One chapter deals with letters, instructions, and short business forms. Discussion of all topics, whether
it is the use of tact in editing another's work or the proper placement of pronouns, is sophisticated.

753 Timm, Paul R. Managerial Communication: A Finger on the
      Pulse. Englewood Cliffs, N. J. : Prentice-Hall, 1980.

754 Tracy, R. C. , and Jennings, H. L. Handbook for Technical
      Writers. Chicago: American Technical Society, 1961.
      Index. Bib. 134p.
This book is geared to professionals who prepare government reports
either from an agency or contractor. The five chapters include "The
Technical Writing Function, " "Elements of Technical Documents, "
"Technical Writing Style, " "Technical Writing Mechanics, " and "Security Requirements. " A 21-page glossary of technical terms and
an 18-page list of general abbreviations are also included. The book
is set in typewriter font.

755 Tracy, Raymond C. , and Jennings, Harold L. Writing for In-
      dustry. Chicago: American Technical Society, 1973.
      Index. 280p.
This book is a guide for technical students. The 11 chapters include
"Building Sentences, " "Descriptions and Instructions, " "Planning the
Report, " and "Forms for Specific Reports. " Each chapter ends with
writing exercises and discussion questions. Appendix A is a "Review
of Grammar"; Appendix B is "Punctuation"; Appendix C is "Capitalization"; and Appendix D is a "Glossary of Vocational Writing Terms. "

756 Treece, Malra. Communication for Business and the Profes-
      sions. Boston: Allyn and Bacon, 1978. Index. 603p.
This text of 20 chapters and its accompanying 201-page teacher's
manual are designed to cover most aspects of a full business
communications course (one, two, or even three semesters). Part
I, "An Overview of Communication, " covers the communication process, organizational communication, semantics, interpersonal communication, and listening. Part II, "Basic Principles of Effective
Communication, " reviews building goodwill, writing style, and correctness. Part III, "Frequently Written Business Messages, " divides
writing into purpose-related topics: the routine and favorable, the
unpleasant and the uncertain, the sales of products and services, and
credit and collection. Part IV, "Communicating About Employment, "
discusses both career planning and job search. Part V, "Communicating Through Reports, " covers writing, researching, evaluating,

and designing the report.   Part VI, "Communication and Effective
Business Management, " is on communication efficiency and oral reporting.   The four appendixes cover format, usage, and concepts of
effective business communication.   The exercises include cases, discussion questions, writing assignments, and sentence corrections.

757 Treece, Malra.   Successful Business Writing.   Boston: Allyn
        and Bacon, 1980.   Index.   Bib.   412p.
This textbook of 14 chapters in five parts is based on the author's
longer and more comprehensive textbook, Communication for Business and the Professions.   The emphasis is placed on explanation
and illustration of the common business forms: memo, letter, and
report.   Parts I and II deal with business writing principles and
readability concerns; Part III deals with types of messages: goodwill, favorable, or bad news.   Part IV covers the job search.   Part
V treats business reports.   Exercises include writing assignments
and cases.

758 Tregoe, James H. , and Whyte, John.   Effective Collection
        Letters.   New York: Prentice-Hall, 1924.   514p.
This book was written to help "the business public" overcome "the
problems of collections. "   The introduction begins with a discussion
of good psychology in collection letters, including the "you attitude"
and the use of "straightforward English. "   Part I covers wholesale
collection letters, and Part II, retail collection letters.   Part III illustrates exchanges (between creditors and customers) and provides
criticism of those exchanges.   Part IV illustrates collection devices,
including "stunt" and "humorous" letters.   Part V gives information
on sequences for collection letters.   Part VI shows poor collection
letters, giving a brief commentary on each letter.   The samples
used throughout the book are taken from actual letters of businesses.

759 Trelease, Sam F.   How to Write Scientific and Technical Pa-
        pers.   Cambridge, Mass.:   MIT Press, 1969.   Index.
        Bib.   185p.
Originally published by Williams and Wilkens in 1958, this is a combination of two earlier books: Preparation of Scientific and Technical Papers (1925, 1927, 1936) and The Scientific Paper, How to
Write It (1947, 1951).   It is a manual for students or researchers
preparing papers or reports on scientific subjects.   The emphasis
is on preparation of theses or dissertations.   Topics covered in the
seven chapters include "The Research Problem, " "Writing the Paper, " "Good Form and Usage, " "Tables, " "Illustrations, " "Prepubli-
cation Review, " and "Proofreading. "

760 Tressler, Jacob C. , and Lipman, Maurice C.   Business Eng-
        lish in Action.   2nd ed.   Boston: Heath, 1957.   Index.
        Bib.   529p.
Although this textbook covers theory, its emphasis is on practice.
The text is divided into two parts with 30 chapters.   Part I, "Speak-
ing and Writing on the Job, " treats various kinds of business letters
in ten chapters and speech in four chapters.   Special topics include
"Personality and Business Relations, " "Telegrams, Cablegrams, and

Memorandums, " and "Postal and Banking Information. " Numerous
examples, pictures, and humorous illustrations are used. The chapter on "Vocabulary of Business" introduces the student to technical
terms/jargon in various professions. Part II is a "Handbook of
Grammar and Usage. " The exercises include fill-in and writing
assignments.

761 Turner, Barry. Effective Technical Writing and Speaking.
        Boston: Cahners, 1974.

762 Turner, Rufus P. Grammar Review for Technical Writers.
            rev. ed. San Francisco: Rinehart, 1971. Index. 118p.
This book of 14 chapters is designed as a supplement to a textbook.
The first chapter discusses language as it relates to thinking and
communication, and the nature, scope, and limitations of grammar.
It also discusses the correct attitude of the technical writer toward
grammar. Chapters 2-12 cover parts of speech and parts of sentences. The chapters are in handbook style, using military numbering. All examples are taken from technical contexts. Brief exercises are also included.

763 Turner, Rufus P. Technical Report Writing. New York: Holt,
            Rinehart and Winston, 1965. Index. 210p.
This text assumes that "technical report writing is an upper-division
course and that the student comes to it fortified by at least one
course in English composition. " The ten chapters cover such topics
as nature, levels, and types of technical literature; formal and informal reports; general procedures for report writing; planning; collecting material; outlining; the rough draft; illustrations; revision;
and reproduction and printing. The chapters end in substantial lists
of writing and practice exercises. The six appendixes offer six
specimen reports: memo report, student report, letter report, two
long professional technical reports, and a laboratory test report.

764 Turner, Rufus P. Technical Writer's and Editor's Stylebook.
            Indianapolis: Sams, 1964. Index. Bib. 208p.
This book is organized both as a stylebook for the professional technical writer and as "a secondary text to be used by the student of a
technical writing course. " Chapter 1 defines technical style. Chap-
ter 2 treats simplifying words and sentences. Chapters 3-10 include
"Capitalization, " "Abbreviation of Common and Technical Terms, "
"Use of Numbers, " "Devices for Emphasis, " documentation techniques, "Symbols, " "Prefixes and Suffixes, " and "Punctuation. "
Chapter 11 covers "Preparing the Manuscript. " Chapter 12 is a
very short handbook of usage with entries alphabetically arranged.
There are six appendixes: "Sentence Faults, " "Proofreader's Marks, "
"Greek Alphabet, " "Symbols for Mathematical Calculations, " "Medi-
cine and Pharmacy Symbols, " and "Astronomy Symbols. "

765 Tuttle, Robert E. , and Brown, C. A. Writing Useful Reports:
            Principles and Applications. New York: Appleton-Century-
            Crofts, 1956. Index. 635p.
The authors of this textbook reject the "types, " "style, " and "freshman

composition" approaches to teaching report writing in favor of one
in which "the report writer analyzes the report situation, ... examines the means [functional elements] at his disposal ... and selects or modifies to fit the situation. " The 17 chapters are divided
into two parts. The first seven chapters are devoted to the potential functions of various structural parts of the report (prefatory
material, introductions, conclusions, etc.) as well as general advice
about the use of reports. The next chapters cover organization
within parts and the use and interpretation of data. After two chapters on using illustrations and the "help of others" (acknowledging
and quoting), a chapter is devoted to the writing process. The final
three chapters in Part I cover format, research, and note taking.
Part II contains a chapter showing how parts of reports can be re-
arranged, a chapter containing sample reports, a job-application letter, and an appendix containing the answers to library research
questions in Chapter 14. A writing checklist follows the index.

766 Ulman, Joseph N. , and Gould, Jay R.   Technical Reporting.
       3rd ed.   New York: Holt, Rinehart and Winston, 1972.
       Index.   Bib.   419p.
According to the authors, this book is written for "students and practitioners of engineering and the sciences who have ... something to
say. " The 20 chapters stress report strategies and forms, writing
style, and grammar. Part I, "Basic Issues, " covers the report as
a sales tool, audience and organization, collection of data and outlining, and technical description. Part II, "The Report, " presents
various types of formal and informal reports, including laboratory
reports, instructions, proposals, and speeches. Part III, "Tools
and Methods, " covers writing style, grammar, mechanics, and illustrations. Chapters include writing assignments, cases, and questions for class discussion. The appendix includes sample technical
reports and letters, in addition to abbreviations for scientific and
engineering terms. Military numbering is used throughout.

767 Uris, Auren.   The Blue Book of Broadminded Business Behav-
        ior.   New York: Crowell, 1977.

768 Uris, Auren.   Memos for Managers.   New York: Crowell,
        1975.   Index.   231p.
This book, set in large type with a 9½ " x 8" format, is designed as
a quick guide for experienced business people. The book is divided
into two parts. Part I has ten sections on topics related to memo
writing, such as when, why, and to whom to write. In addition,
these sections include the political strategy of memos, a "memo-
usage analyzer, " and avoiding "overcommunication. " Part II is
composed of sample memos with short introductions to each type,
titled according to purpose. The types of memos are alphabetically
arranged, beginning with "acknowledgments" and "admonitions" and
ending with "welcome" and "you were right" memos. The sample
memos tend to be breezy and clever.

769 Van Hagen, Charles E.   Report Writer's Handbook.   Englewood
        Cliffs, N.J.: Prentice-Hall, 1961.   Index.   Bib.   276p.

The aim of this handbook is to aid professionals, such as "practicing scientists, engineers, economists, psychologists, and administrators," in preparing papers and reports as a part of their work. The seven chapters deal with the major sections of reports. Each chapter is further subdivided into units. Chapter 3, "Front Matter," discusses "covers," "title page," "letter of transmittal," and the like. The chapters treat all the major facets of report writing: the body, the front matter, the end and supplementary matter (including illustrations). Special "Notes," provide additional guiding principles for solving troublesome problems.

770 Vardaman, George T., and Vardaman, Patricia Black. Communication in Modern Organizations. New York: Wiley, 1973. Index. Bib. 516p.

According to its authors, this "book is designed principally as a four-year or two-year college text, and it is intended to equip the student with both basic communication strategies and tactics for superior results ... emphasizing ... written communication." The 17 chapters are divided into four parts. Part I covers some general principles of communication, letters, memos, reports, legal documents (proposals, agreements, and directives), manuals, forms, and brochures. Part II covers forms of writing for specific purposes: personal, organizational, and professional. Some forms related to these purposes are letters of complaint, policy statements, procedural reports. Also covered are letter design and format. Part III presents the use of communication forms other than writing and how to use each best. The forms range from oral presentation to automated systems. Part IV serves as a reference handbook on language usage, mechanics, and principles of composition. Part V provides specimens of various kinds of writing: letters, reports, ads, memos, policy statements, résumés, etc.

771 Vardaman, George T., and Vardaman, Patricia Black. Successful Writing: A Short Course for Professionals. Wiley Professional Development Programs. New York: Wiley, 1977. 319p.

This publication is composed of 11 programmed booklets of varying length, contained in a cardboard slipcase. This work, designed as a self-study guide for professionals, teaches the principles of writing based on the acronym PRIDE: Purpose, Receiver, Impact, Design, and Execution. Each booklet begins with a statement of objectives that should be fulfilled when the user completes the book. The first booklet provides an overview of the role of the writer and the needs of the audience; the last offers instruction in grammar, vocabulary, and mechanics. The other booklets are grouped under three categories: Purpose, Design, and Execution. "Purpose" (booklets 2, 3, and 4) deals with building goodwill, handling complaints and adjustments, and rejecting proposals or reprimanding delinquent accounts. "Design" (booklets 5, 6, and 7) emphasizes the psychological design of documents and letters. "Execution" (booklets 8, 9, and 10) treats ways of writing letters in four categories: nonroutine, routine with prepared parts, routine but complete, and mass-distribution letters. The programmed exercises require the reader to recall the principles taught earlier in the booklet.

772 Vernon, Cay. Supreme Letter Writer. New York: Scully,
        1928. 226p.
This model-letter book begins with "Ten Golden Rules," a list of
tips on letter writing. Otherwise, the book is a collection of model
letters covering both personal and business situations. There are
ten sections: the five sections covering business letters include
"Personal Business Letters," "Business Letters," "Letters Regarding Real Estate," "Letters About Employment," and "Letters Regarding Banking." A final part, Section 11, explains how to address important persons.

773 Waldo, Willis H. Better Report Writing. 2nd ed. New York:
        Reinhold, 1965.

774 Wales, LaRae H. A Practical Guide to Newsletter Editing and
        Design. 2nd ed. Ames: Iowa State University Press,
        1976. 52p.
According to its author, this "guide is for inexperienced editors ...
who want to put out a high quality newsletter even though they have
a small budget." The eight chapters cover the essential questions
and details an editor must answer and know. Chapter 1, for example, gives a two-page overview of the purposes, audience, and needs
for newsletters. Chapter 2 reviews the editor's duties and the writing style appropriate to newsletters. Chapter 3 covers basic decisions an editor must make, such as choosing a format, selecting
paper, and preparing a mailing list. Chapters 4-6 cover the issues
involved with printing: selecting typeface, choosing a printing method, etc. Chapter 7 covers the use of photography, and Chapter 8,
design elements. A glossary of technical terms is included.

775 Walker, Charles Francis, and Robertson, Mary. Practical
        Business Correspondence for Colleges. 3rd ed. Cin-
        cinnati: South-Western, 1966.

776 Walter, John A. Report on Technical Report Form. rev. ed.
        Austin, Tex.: University Co-operative Society, 1973.

777 Walton, Thomas F. Technical Manual Writing and Administra-
        tion. New York: McGraw-Hill, 1968. Index. Bib.
        383p.
This book of 13 chapters emphasizes the preparation of military manuals and gives a general analysis of manual production. It begins
with an overview of the types and purposes of technical manuals and
explains the manual-production process from applying technical-
manual specifications to planning the research phase and writing
various portions. Special types of manuals, such as procedural and
special-purpose, as well as the validation and verification of manuals are discussed. A section is also included on editing, along
with sections on such topics as "Application of Automation and Audio-
Visual Aids in Communication," "Management of Technical Publication Programs," and "Preparing Proposals and Contracts."

778 Ward, Ritchie R. Practical Technical Writing. New York:

Knopf, 1968. Index. Bib. 264p.
This textbook, written for students in the natural sciences, has ten chapters divided into four parts. Part I, "Your Reader," covers audience analysis, purpose, clarity, affectation, and Gunning's Fog Index. Part II, "Your Purpose," covers determining the scope of a report, outlining, and the paragraph. Part III, "Your Subject," treats sentence patterns, grammatical problems, sentence variety, word choice, and final editing. Included at the end of Part III is an article, "The Principles of Poor Writing," by Paul Merrill. Part IV is the appendix, which contains three samples of technical writing: an abstract, a book chapter, and a technical report. The exercises throughout are basic writing assignments.

779 Warren, Thomas L. Technical Communication: An Outline.
         Totowa, N. J. : Littlefield, Adams, 1978. Index. Bib.
         148p.
This book of five chapters is written totally in outline form, with the exception of the appendixes. According to the author, the reader is a "technical communications beginner"--but presumably not a student, since the book contains no exercises. The book's purpose is "to provide in as simplified form as possible the fundamentals of technical communication." In addition to an introductory section, the book outlines prewriting, writing techniques for various purposes, postwriting (revision and clarity checks), and technical writing forms (including oral reports). The six appendixes cover clarity, abbreviations, words and phrases to avoid, a sample personal data sheet, a letter of application, and a bibliography.

780 Watson, Herbert. Applied Business Correspondence. Chicago:
         Shaw, 1922.

781 Watt, Homer A. , and McDonald, Philip B. The Composition
         of Technical Papers. 2nd ed. London: McGraw-Hill,
         1925. Index. 429p.
The first edition of this textbook, published in 1917, was written entirely by Watt; the revision was prepared by McDonald, who wrote two new chapters: "Diction and Professional Style" and "The Campaign for Better Letters." The text is designed to enable engineering students to grasp the general principles of expository writing and then apply them to their specific task of technical exposition. The 12 chapters are divided into two parts. Part I is devoted to the planning and preparation of the whole composition, from sentences to the whole. Part II deals with such topics as types of exposition (process, description, etc.), "Engineer's Reports," and "Business and Professional Letters." Although numerous samples of reports and illustrations are included, the book contains no exercises or other pedagogical apparatus. Student examples are included.

782 Wattles, Gordon H. Writing Legal Descriptions. Santa Ana,
         Calif. : Wattles, 1976. Index. 315p.
This textbook deals with describing land surveys, leases, and titles in legal language. The 14 chapters deal with writing legal points of land descriptions while maintaining clarity. Three chapters

are specifically devoted to writing: Chapter 2, "Finding Support Information"; Chapter 3, "Description Fundamentals"; and
Chapter 11, "Writing Descriptions." The book contains problems for discussion and writing exercises at the end of each
chapter.

783 Weaver, Patricia C., and Weaver, Robert G. Persuasive
     Writing: A Manager's Guide to Effective Letters and
     Reports. New York: Free Press, 1977.

784 Webster, E. H. English for Business. New York: Newson,
     1916.

785 Weeks, Francis W., ed. Readings in Communication from
     Fortune. New York: Holt, Rinehart and Winston, 1961.
     143p.
Each of the 28 articles from Fortune magazine, ranging from topics
like "What We Know About the Process of Communication" to "Pragmatic Communicators," begins with an informative abstract. Articles
include "Why Professors are Suspicious of Business," by Bernard
DeVoto; "Says Business to Mr. DeVoto," by Albert Lynd; and "To-
morrow's Telephone System," by Francis Bellow. The editor states
that the purpose of this collection of readings is to provide a "broad
panorama of the problems of communications and the attempted solutions that will supplement textbooks in Management, Human Relations
in Business, Mass Communications, Public Relations, Speech, Business Communication, ... and Advertising."

786/7 Weeks, Francis W., and Hatch, Richard A. Business Writing
     Cases and Problems. Champaign, Ill.: Stipes, 1977.
     116p.
The aim of this casebook is to provide "practice not only in business
writing and speaking, but also in thinking about and coping with business situations ... both routine and complex...." Therefore, the
cases often lack nonessential information so that students can invent
details to help resolve the hypothetical situations. Since the book is
designed as a supplement for a basic business communication text,
the authors provide only a brief discussion of the principles to be
used in responding to the cases. The subjects covered include writing informative, persuasive, and negative messages; writing job applications; and developing a business writing style. In addition, a
brief handbook treatment is given to letter format, punctuation,
numbers, and grading standards and correction symbols.

788 Weeks, Francis W., and Jameson, Daphne A. Principles of
     Business Communication. 2nd ed. Champaign, Ill.:
     Stipes, 1979. Bib. 156p.
As in the first edition by Francis W. Weeks, the stated purpose of
this book is "to provide a foundation for the study and discussion of
problems in business communication." It is not intended to be the
sole text for a business communication course; in fact, a list of
texts that do contain exercises and other classroom material is given

at the end of each chapter. The 12 chapters are written in a chatty
style. The first two chapters discuss general principles of business
communication theory and rhetoric. Throughout the remaining ten
chapters of the book, the authors cover the "whys" rather than the
"hows" of topics in business communication, such as semantics,
visual details, writing style, résumés and application letters, and
persuasive writing.

789   Weihofen, Henry. Legal Writing Style. St. Paul, Minn.:
          West, 1961.

790   Weil, Benjamin H. Technical Editing. Westport, Conn.:
          Greenwood, 1975.

791   Weil, Benjamin H., ed. The Technical Report: Its Prepara-
          tion, Processing, and Use in Industry and Government.
          New York: Reinhold, 1954. Index. Bib. 485p.
This collection of 24 articles is for professionals who must write
reports and for publication departments in industry and government.
There are five parts. Part I, "Functions," includes articles about
the purposes of reports in industry and government. Part II, "Preparing and Processing," treats style, editing, and writing various
types of technical reports. Part III, "Distributing," discusses distributing industrial reports and ensuring the security of government
reports. Part IV, "Filing," deals with indexing, cataloging, and
abstracting reports; Part V, "Using," covers locating and utilizing
reports for various purposes. The two appendixes provide a report
manual and a reprint of Executive Order 10501 on the safeguarding
of classified information.

792   Weiner, Solomon. Business Letter Writing. New York: Mon-
          arch, n.d.

793   Weisman, Herman M. Basic Technical Writing. 4th ed. Co-
          lumbus, Ohio: Merrill, 1980. Index. Bib. 414p.
This textbook of 17 chapters is divided into five parts. Part I, "Introduction," contains chapters on the history and theory of technical
writing, semantics and the communication process, and technical
writing style. Part II, "Fundamentals," covers basic expository
techniques: definition, description, explaining a process, and analysis. Part III, "Technical Report Writing," explains the scientific
method, investigation, organization, format, writing, and illustrating.
Part IV, "Shorter Technical Forms," treats papers and professional
articles, technical correspondence, proposals, and oral reports.
Part V consists of an alphabetical index of usage, punctuation, and
grammar, in addition to exercises and problems in grammar and
usage. The two appendixes give requirements for progress and final
reports and examples of technical writing forms covered in the text.
Discussion questions and writing assignments are at the ends of the
chapters.

794   Weisman, Herman M. Technical Correspondence: A Handbook

and Reference Source for the Technical Professional.
New York: Wiley, 1968. Index. Bib. 218p.
This book is designed for scientists and engineers "who meet their
communication requirements with difficulty, aversion, and, often,
hostility. " It is, therefore, written in a personal, yet instructive,
style.   The ten chapters are in three parts.   Part I, "Principles
and Fundamentals, " discusses the communication process, the psychological and stylistic principles of technical correspondence, planning letters, and the format of correspondence.   Part II, "Applying
Correspondence Principles, " covers various types of correspondence:
inquiries and requests, sales and proposals, employment letters,
memoranda, and other professional, and even personal, letters.
Part III is the appendix, which contains a 20-page, alphabetical
grammar, punctuation, and usage guide.

795 Weisman, Herman M.   Technical Report Writing.   2nd ed.
          Columbus, Ohio: Merrill, 1975. Index. Bib. 181p.
Abridged from Basic Technical Writing, by the same author, this
book focuses only on technical reports.   It begins with a chapter
titled "What Technical Writing Is, " which not only defines technical
writing but also gives historical details and the special problems of
"factual" communication.   Chapter 2 discusses semantics and the
process of communication and the role of these fields on the scientific method and technical writing.   Chapters 3-5 focus on the report
itself--researching, organizing, and writing.   Chapter 6 covers graphic presentation of information; Chapter 7 discusses technical style;
e. g., clarity, conciseness, objectivity, and precision.   Following
the chapters is a 26-page, alphabetically arranged reference guide
to grammar, punctuation, style, and usage.

796 Weiss, Allen.   Write What You Mean: A Handbook of Business
          Communication.   New York: AMACOM, 1977.   Index.
          179p.
This book is intended for those who must prepare memoranda, reports, and speeches.   It also briefly covers professional and public-
relations writing in business.   The 13 chapters are divided into four
parts.   Part I covers preliminary concerns, such as writer/reader
attitudes, and examines types of writing (e. g., letters, memos,
manuals) as options from which to choose.   Part II discusses information gathering and outlining.   Part III covers openings, methods
of development, paragraphing, and clarity.   Part IV treats usage,
style, the purpose of grammar, professional writing in business,
and business speaking.   The book stresses advantages and disad-
vantages of various formats.

797 Wellborn, G. P.; Green, L. B.; and Nall, K. A.   Technical
          Writing.   Boston: Houghton Mifflin, 1961. Index. 374p.
The ten chapters of this textbook are divided into two parts.   Part
I, "Principles and Techniques, " covers the definition of technical
writing, research methods, organization, definition and description,
and graphics.   Part II, "Forms of Technical Communication, " covers
business communications, summaries, reports, articles, and speeches.
Following the chapters, which have writing and class-discussion ex-

ercises, are five appendixes, which provide samples of letters, abstracts, reports, articles, and speeches.   All the examples and exercises assume a science or engineering background.

798 Wells, Walter. Communications in Business. 2nd ed. Bel-
        mont, Calif.: Wadsworth, 1977. Index. 528p.
This textbook emphasizes the fundamentals of business writing and
the "behavioral distinctions" involved in business writing, such as
requests versus demands, persuasion versus conciliation.   The 18
chapters are divided into five parts.   Part I, "The Basics of Business Writing," covers writing style.   Part II, "The Fundamentals
of Business Letters and Memoranda," covers effective business tone,
format, and routine letters.   Part III, "Letters and Memos: The
More Difficult Kinds," deals with persuasion, good news and bad
news, job applications (including job changing), and demands.   Part
IV, "Report Writing in Business," discusses the varieties of reports,
organization, style, and format.   Part V, "The Art of Speaking and
Listening in Business," covers telephone techniques and face-to-face
communication.   There are six appendixes, covering forms of address, international business letters, common abbreviations, a glossary of business terms, grammar and punctuation, and a list of correction symbols for teachers and students.

799 Weseen, Maurice H. Write Better Business Letters. New
        York: Crowell, 1933.

800 West, M., and Kimber, P. F. A Deskbook of Correct Eng-
        lish. London: Longmans, Green, 1958.

801 Whalen, Doris H. Handbook for Business Writers. New York:
        Harcourt Brace Jovanovich, 1978. Index. Bib. 269p.
This book has two purposes: it is "intended to serve as a textbook
and training manual for the student or beginning writer" and to serve
as a "ready reference manual" for language skills and business
writing forms for the experienced business writer.   The only pedagogical apparatus is the 24-page exercise section at the back of the
book.   Its ten sections include general points, such as discussions
of the business letter, good business writing, language choice, and
more specific topics, such as business letters for specific purposes,
employment writing, business-report writing, personal business writing, punctuation, and style.   A glossary of grammar and usage, exercises, and a key to exercises follow the main sections.   This
handbook also contains material useful to secretaries from The Secretary's Handbook, by the same author.

802 Whalen, Doris H. Handbook of Business English. New York:
        Harcourt Brace Jovanovich, 1980. Index. 264p.
This handbook has a dual purpose as a textbook for students who
need fundamental language skills and "as a reference manual of acceptable English usage" for those on the job.   The 21 sections cover
such topics as the sentence, parts of speech, punctuation, spelling,
and usage.   The 80-page exercise section includes an answer key.
Also included is a glossary of grammatical terms and appendixes on

abbreviations, American and Canadian postal information, the metric
system, and the format of a business letter.

803   Whitburn, Merrill, ed.   Teaching Technical Writing: The First
      Day in the Technical Writing Course.   Morehead, Ky.:
      Association of Teachers of Technical Writing, 1979.   25p.
This collection of five articles is part of an anthology series designed
to guide teachers in technical writing.   Selections include "The Letter
of Inquiry as an Opening Exercise, " by Stephen Gresham and William
Rivers; "Beginning Where Careers Begin: The Resume, " by Gordon
E. Coggshall; and "Comparative Analysis and Audience Levels, " by
Wayne A. Losano.   "An Important Discovery: Purpose Determines
Language, " by Ron Dulek, explains opening the class by analyzing
a poem and then progressing to a discussion of language adaptation.

804   White, Virginia P.   Grants: How to Find Out About Them and
      What to Do Next. New York: Plenum, 1975. Index.   354p.
This study of the process of getting grants is divided into four major
sections: "What Is a Grant?, " "How to Find Out About Grants and
Who Gives Them, " "The Application, " and "Grantsmanship. "   These
sections contain chapters that deal with most of the facets of the
grant process.   For example, some of the chapters deal with library
research, workshops and institutes, government grants, foundations,
and writing the proposal.   The section on writing the proposal covers
selecting the agency, determining costs, patents and copyrights, the
format of the proposal, and credentials of the staff.   The chapter
on how grants are awarded offers detailed background for the would-
be grant writer.   Eight appendixes follow the chapters and include:
(1) a list of types of grants and definitions, (2) a list of abbreviations of federal agencies, (3) a list of HEW offices and addresses,
(4) a list of Public Health Service programs, (5) a list of Federal
Information Centers and telephone tielines (toll free), (6) a list of
libraries with collections of regional foundation information, (7) a
list of HEW regional comptroller's offices, and (8) a U.S. list of
endangered fauna.

805   Whittem, A. F., and Andrade, M. J.   Spanish Commercial
      Correspondence.   Boston: Heath, 1916.

806   Whyte, William H., Jr., et al.   Is Anybody Listening?   New
      York: Simon and Schuster, 1952.   239p.
This chatty yet thoughtful book for business professionals about communication and writing in business deals with the attitudes that produce poor communication in business.   The author admits that he
and his coauthors (the editors of Fortune) are not communication
experts, but they do have useful observations about the failure of
American business to talk to Americans.   Throughout the 11 chapters, the authors use cartoons and provocative comments to communicate with their readers (professionals in business who are also not
communication experts).   For example, in Chapter 4, "The Prose
Engineers, " they conclude their assessment of the value of the readability experts by asserting, "Thus the readability movement is also
the measure of a failure in our schools and colleges. "

807   Wicker, Cecil V. , and Albrecht, W. P.   The American Tech-
      nical Writer: A Handbook of Objective Writing.   New
      York: American Book, 1960.   Index.   Bib.   415p.
According to the authors, the 17 chapters of this textbook emphasize
"rhetorical principles and illustrates their application in various forms
of professional writing. "   The book uses the term "objective" rather
than "technical" writing because it emphasizes expository writing.
For example, Part I, "Principles of Technical Writing, " devotes 11
chapters to such subjects as the principles of objective exposition;
elements of discourse; audience adaptation; forms of discourse (including analysis of mechanisms and processes); interpretation; evaluations, standards, and specifications; research; and the method of
science.   Part II, "Forms of Technical Writing, " covers abstracts,
reviews, letters, reports, articles, and speeches.   Part III, "Handbook, " serves as a student reference for grammar, style, usage,
and manuscript preparation.   All sections contain exercises for writing and class discussion.

808   Wiener, Solomon.   Business Letter Writing.   New York: Mon-
      arch, 1973.

809   Wilcox, Roger P.   Communication at Work: Writing and Speak-
      ing.   Boston: Houghton Mifflin, 1977.   Index.   481p.
This textbook, based on the author's consulting experience at General
Motors, is geared for both students and those on the job.   The 12
chapters include the basics of grammar, punctuation, usage, and effective sentence construction.   Part I, "Overview, " is devoted to the
communication process.   Part II, "Being Clear, " deals with organization of business writing, effective paragraphing, and graphic aids.
Part III deals with persuasion, and Part IV with specific applications
of communication techniques to forms of business writing, such as
letters, memos, reports, and proposals.   Part V treats oral communication.   The exercises include writing assignments, questions
for class discussion, and cases.

810   Wilcox, Sidney W.   Technical Communication.   Scranton, Pa. :
      International Textbook, 1962.   Index.   Bib.   306p.
This textbook of 13 chapters is designed for the upper-level engineering student; the examples used assume some knowledge of engineering
terms and the engineering method.   Chapter 1 covers the general issues of language and technical writing style.   Chapters 2-4 treat the
modes of discourse appropriate to technical writing: Definition, Description, and Narration (Process Description).   Chapters 5-7 cover
methods of development (analysis and argument), coherence (introductions, transition, conclusions), and writing style.   Chapters 8-10
treat several forms of technical writing: the short report, the research paper, and articles for publication.   Chapter 11 discusses
public speaking, and Chapter 12 treats business communication as
it relates to technical people.   Chapter 13 provides a four-year reading list to develop "Professional Literacy. "   All the chapters contain
writing exercises or discussion questions, or both.   Two appendixes,
covering technical abbreviations and 107 commonly misspelled words,
are included.

811   Wilkinson, C. W. ; Menning, J. H. ; and Anderson, C. R.
      Writing for Business: Selected Articles on Business
      Communication.   3rd ed.   Homewood, Ill. : Irwin, 1960.
      369p.
This collection of 78 articles by such authors as Rudolf Flesch and
William Butterfield and from such publications as Printer's Ink and
Business Week is divided into ten parts.   The selections cover letter
writing principles, reader adaptation, good style, readability, reports, and types of letters (claims/adjustments, credit/collection,
inquiry/acknowledgments, sales, and application).   Each article contains a brief biographical sketch of its author.

812   Williams, Cecil B. , and Griffin, E. Glen.   Effective Business
      Communication.   3rd ed.   New York: Ronald, 1966. In-
      dex.   580p.
The authors of this third edition (previously titled Effective Business
Writing by Cecil B. Williams and John Ball) have expanded their
treatment of writing and have "added a chapter on general communication, dealing with reading, speaking, and listening as they pertain
to business. "   The textbook, designed for college and university students, contains writing and class-discussion exercises as well as
end-of-chapter readings.   The 17 chapters are divided into four
parts: "The Fundamentals, " "Business Letters, " "Other Business
Writing, " and "Appendix. "   The first part deals with the psychology
of business communication and the nature of sentences and vocabulary
used in business writing; the second part, types of business letters;
and the third, reports and general communication matters, such as
direct-mail advertising and dictation.   The appendixes include reference sections for postal information, business terms, and letters
and the law.

813   Williams, George E.   Technical Literature: Its Preparation
      and Presentation.   London: Allen and Unwin, 1948.
      Index.   Bib.   117p.
This work is written primarily for engineers and physicists to help
them "in preparing technical and scientific papers for the professional press. "   The author also suggests that the book will be of assistance to "technical writers in research associations, in manufacturing and operating companies, and in the Services. "   In the first
of seven chapters, the author presents the argument that technical
writing can be considered as a literary form: "If the language is
clear and convincing, if all the parts of the subject-matter are interrelated to form an organic whole which is something more than
the mere sum of the parts, the product may rank as literature; if,
in addition, the information is accurate and the argument logical, it
may properly be described as 'technical literature. '"   The author
continues in Chapter 2, "Method of Presentation, " with a discussion
of logical arrangement and sequence in addition to writing the first
draft.   Chapter 3, "The Choice of Words, " covers various facets
of style: use of passive voice, jargon, technical terms, and language in general.   Chapter 4, "Organization, " treats arrangement
of sentences, whole parts of reports, and the numbering of sections.
Chapter 5, "The Preparation of Manuscripts, " covers the use of

punctuation, symbols, layout, and editing techniques. Chapter 6, "The Art of Sub-Editing," treats editorical practices after the work has left the author. Chapter 7, "illustration," gives an 11-page guide to graphics. The two appendixes are an essay, "Psychological Principles," designed to help the writer meet the reader's needs, and a brief style guide titled "Standards."

814 Wilson, George Frederick. House Organ--How to Make It Pro-
          duce Results. Milwaukee: Washington Park, 1915. 199p.
This book is designed for those who must edit house organs. The first of 15 chapters treats the history and purpose of house organs. Chapter 2 covers basic questions about editorial policy. Chapters 3-11 cover the stages in preparing a house organ, from determining who will receive it and what physical size it should be to the use of illustrations and editing copy. Chapters 12-13 treat the internal and external house organ. Chapters 14-15 give advice on mechanical and editorial details the editor must face. A number of photographs of house-organ covers are included as well as sample content from current house organs.

815 Winfrey, Robley. Technical and Business Report Preparation.
          3rd ed. Ames: Iowa State University Press, 1962. In-
          dex. Bib. 340p.
This book is aimed at helping those in science, technology, and government "acquire skill and effectiveness in the preparation of their reports." Among the 17 chapters are treatments of "Planning the Investigation and Report," "Collection of Information," "Prepara-tion of Illustrations," "Format and Arrangement," "Oral Presenta-tion of Reports and Technical Papers," and "Copying and Duplicat-ing." Chapter 1, "Reports," explains the purposes and types of reports. Chapter 2, "Correspondence," covers letters and memos because, according to the author, considerable correspondence may take place before a report is written. The book is profusely illus-trated with samples of formats in all phases of the report process: bibliography, questionnaire, charts, letters, etc. Two appendixes, "References for Collateral Reading" and "Specimen Formal Report--Relocating the Union Bus Station in Ames, Iowa," provide further assistance.

816 Wirkus, Tom E., and Erickson, Harold P. Communication
          and the Technical Man. Englewood Cliffs, N.J.: Prentice-
          Hall, 1972. Index. 237p.
This textbook, "written on the thirteenth-to-fourteenth grade read-ability level," is geared to the written and oral communication needs of students in English, technical-writing, speech, and general communication courses. The four chapters provide in-depth treat-ment of such topics as "Written Technical Communication" and "Oral" Communication." The chapters on these topics reinforce each other through common concepts, such as audience analysis, planning, and preparation. There are exercises in each chapter for writing and class discussion. Appendixes include a brief handbook of usage and sample technical reports.

817   Woelfle, Robert M. , ed.   A Guide to Better Technical Presen-
        tations.   New York: IEEE, 1975.   Index.   230p.
This collection of papers and articles is a part of the "Selected Reprint Series" of the Institute of Electrical and Electronic Engineers.
It contains 30 selections about preparing, writing, editing, illustrating, and presenting a technical paper.   Many of the papers in the
collection have appeared in Technical Communication or have been
presented at Society for Technical Communication meetings.

818   Wolf, Morris Philip; Keyser, Dale F. ; and Aurner, Robert R.
        Effective Communication in Business.   7th ed.   Cincin-
        nati: South-Western, 1979.   Index.   515p.
This textbook has 19 chapters divided into nine parts.   The first
seven parts deal with communication theory and goals, communication planning, multipurpose business messages, career planning and
job promotions, persuasive communication, bad-news messages, and
reports.   Part VIII provides cases, "minidramas" and "scenerios"
for class discussion.   Part IX is a handbook of usage and grammar.
Each chapter ends in exercises for discussion and writing.   Each
chapter also ends in "Review and Transition, " a section that summarizes the chapter and introduces the next one.

819   Woodford, F. Peter, ed.   Scientific Writing for Graduate Stu-
        dents: A Manual on the Teaching of Scientific Writing.
        New York: Rockefeller University Press, 1968.   Index.
        Bib.   190p.
This teacher's guide is designed to offer as much practical help as
possible to reduce the teacher's load.   At the beginning of many of
the 14 chapters, for example, the teacher is given a list of materials needed, readings, an estimate of the class time the subject
will require, and possible student assignments.   The teacher is also
given detailed commentary on the in-class projects suggested.   The
first nine chapters (by Woodford himself) cover the steps in writing
and publishing a journal article--from planning to outlining to responding to an editor's comments.   The final five chapters (by other
authors) cover related topics: "Design of Tables and Figures, "
"Preparation for Writing the Doctoral Thesis, " "Writing a Research
Proposal, " "Oral Presentation of a Scientific Paper, " and "Principles and Practices in Searching the Scientific Literature. "

820   Woolcott, Lysbeth A. , and Unwin, Wendy R.   Communication
        for Business and Secretarial Students.   London: Macmil-
        lan, 1974.

821   Wright, Waldo C.   Business Correspondence.   Indianapolis:
        Bobbs-Merrill Educational, 1967.   195p.
This high school textbook has 15 units covering such topics as
"Choosing Correct Words, " "Coherence Through Paragraphing, "
"Sales Promotion Letters, " "Credit and Collection Letters, " and
"Letters of Application. "   The last unit, "Special Reports and Min-
utes, " gives a very brief discussion of types of reports, examples
of simplified wording, and how to take minutes.   The text empha-
sizes that most reports take the letter format.   The units end with

sentence exercises, writing assignments, cases, and review questions.
Unit 8 includes tips for sending telegrams, making reservations, acknowledgments, and night letters.   Grammar units avoid grammar
terminology--e. g. , "putting like forms together" for parallelism.
Each unit begins with a quotation from famous persons (who are
also pictured), such as Henry Ford, Mark Twain, George Eliot, and
Woodrow Wilson.

822   Writing and Publishing Your Technical Book.   New York:
          Dodge, 1959.   Index.   50p.
The purpose of this work is to guide a prospective author through
a book's conception to obtaining support and backing of a publisher.
The six chapters include "Should You Write a Book?, " "Planning
for Success, " "Choosing a Publisher, " "The Prospectus and Outline, "
"Writing the Specimen Chapters, " and "Acceptance and Contract. "
Throughout the chapters are checklists and helpful tips.   In the discussion of analyzing the market, the book advises authors to try to
picture someone they know reading his pages, "Would Old Joe understand this?"

823   Writing Better Letters, Reports, and Memos.   New York:
          American Management Association, 1975.

824   Writing Reports That Work: A Programmed Instruction Course
          for Management Education.   New York: American Man-
          agement Association, 1969.

825   Wyld, Lionel D.   Preparing Effective Reports.   New York:
          Odyssey, 1967.   Bib.   198p.
Although designed primarily as a handbook, this book's nine chapters
do, for the most part, contain discussion questions, problems, or
writing assignments.   The first four chapters deal with the preliminaries (audience and purpose), semantics, style, and the paragraph.
Chapter 5 covers documenting the report.   Chapters 7-8 treat the
short report and the long (or formal) report.   Chapter 9 presents
types of reports that promote the operations of an organization and
the role of an editor.   Four appendixes include "Notes on Business
Correspondence, " "Writing Essentials and Manuscript Mechanics, "
"USAECOM Standards for Technical Reports, " and "Sample Reports. "

826   Wylder, Robert C. , and Johnson, Joan Grissberg.   Writing
          Practical English.   New York: Macmillan, 1966.

827   Yeck, John D. , and Maguire, John T.   Planning and Creating
          Better Direct Mail.   New York: McGraw-Hill, 1961.

828   Zall, Paul M.   Elements of Technical Report Writing.   New
          York: Harper and Brothers, 1962.   Index.   Bib.   208p.
This book of nine chapters begins by presenting the definition and
purpose of reports.   Chapter 2, "Planning, " reviews purpose, audience, and preparing a "target" (thesis) statement.   Chapter 3, "Collecting Information, " covers recording data, using tables and charts,
library research, interviewing, and the questionnaire.   Chapter 4,

"Designing," discusses outlining and format; Chapters 5-6, "Rough Drafting: Introduction and Discussion" and "Rough Drafting: Analysis and Conclusion," treat definition, description, introductions, methods of development, logic, and conclusions. Chapters 7-8 treat revision of sentences and paragraphs and revision for word choice. Chapter 8 also serves as a usage handbook with exercises. Chapter 9, "Revising: Mechanics," serves as a handbook for punctuation, abbreviation, footnotes, pagination, and use of numbers. The chapters contain writing assignments and questions for class discussion.

829   Zetler, Robert L., and Crouch, George W. Advanced Writing. 3rd ed. New York: Ronald, 1961.

830   Zetler, Robert L., and Crouch, George W. Successful Communication in Science and Industry: Writing, Reading, and Speaking. New York: McGraw-Hill, 1961. Index. 290p.

This textbook is designed for students in science and engineering who must "know how to transmit ... ideas to [their] fellows." The 18 chapters are divided into three parts. Part I, "Technical Writing," covers organization; rhetorical methods; letters and reports; and grammar, punctuation, and spelling. Part II, "Reading Techniques," treats word and number recognition, "critical" reading, methods of skimming, and other related topics. Part III, "Speaking Techniques," covers some topics for public speech and group discussion. The chapters contain exercises for writing, revision, and class discussion.

831   Zook, Lola M. Technical Editing, Principles and Practices. Washington, D.C.: Society for Technical Communication, n. d.

RELATED WORKS

The following lists are by-products of our research in collecting titles for the annotated entries. Further, the works that follow are related to the annotated works but are peripheral in one of two ways: either they fall outside our scope as it is stated in the preface or they are so severely limited in their publication (e. g. , industrial style manuals) that they are unavailable to our readers through standard means. We have included these titles, however, for those who might wish to trace them using the information we have provided. We have also provided these lists to help scholars and others who wish to begin study in related fields, such as graphics, oral communication, and publishing.

The lists are not, however, intended to be complete. We have gathered commonly cited titles from other bibliographies and lists. We have not looked at the works themselves; therefore, we cannot vouch for the accuracy of these citations.

Industry and Society Style Guides

American Association of Agricultural College Editors. Communications Handbook. 3rd ed. Danville, Ill.: Interstate, 1976.

American Medical Association. Style Book: Editorial Manual of the AMA. Acton, Mass.: Publishing Sciences Group, 1976.

American National Standard Guidelines for Format and Publication of Scientific and Technical Reports. New York: American National Standards Institute, 1974.

Anderson, Dorothy L. Bibliographic References: A Manual for Engineering Reporters. University Park: Pennsylvania State University Press, 1968.

Beck, L. W. , and Shaefer, Phyllis K. The Preparation of Reports. 3rd ed. Wilmington, Del.: Hercules Powder Company, 1945.

Birth of an Article: Step-by-Step Account with Exhibits Showing How an Article Was Prepared and Processed for Product Engineering. New York: Committee on Education, American Business Press, 1966.

Brennen, Lawrence David. Management Writing Guide. Waterford,

Conn.: National Foremen's Institute, Bureau of Business Practice, National Sales Development Institute, 1963.

Brusaw, Charles T.; Alred, Gerald J.; and Oliu, Walter E. NCR Handbook for Effective Writing. Dayton, Ohio: NCR Corporation, 1974.

CBE Style Manual. 4th ed. Washington, D. C.: American Institute of Biological Sciences, 1978.

Communications Downward and Upward. New York: National Retail Merchants Association, 1961.

Connolly, J. Effective Technical Presentations. St. Paul, Minn.: 3M Business Press, 1968.

DeMare, George. A Handbook of Model Reports to Clients. New York: Price Waterhouse, 1964.

Emerson, Lynn A. How to Prepare Instruction Manuals. New York: State Department of Education, 1950.

_____. How to Prepare Training Manuals. Albany: State Education Department, University of New York, 1952.

Engineering Specifications: Writing Guide. N. p.: Bell Helicopter, Division of Textron, n. d.

Federal Electric Corporation, Training Branch. How to Write Effective Reports. Reading, Mass.: Addison-Wesley, 1965.

Gaddy, L. Editorial Guide. Denver: Martin Company, 1958.

General Notes on the Preparation of Scientific Papers. London: Royal Society, 1974.

Goodman, S. J. A Guide to Instruction Book Preparation. Boonton, N. J.: Aircraft Radio Corporation, 1958.

Haemer, K. W. Writing for the Reader. New York: Business Research Division of AT&T, 1965.

Handbook and Style Manual for ASA, CSSA, and SSSA Publications. Madison, Wis.: American Society of Agronomy, Crop Science Society of America, Soil Science Society of America, 1976.

Handbook for Authors. Bellaire, Tex.: American Journal of Medical Technology, 1975.

Handbook for Authors of Papers in American Chemical Society Publications. Washington, D. C.: American Chemical Society, 1978.

Handbook for Chemical Society Authors. London: Chemical Society, 1960.

Hawkridge, David G., and Campean, Peggie L. Developing a Guide for Authors of Evaluation Reports of Educational Programs: Final Report. Palo Alto, Calif.: American Institute for Research in Behavioral Sciences, 1969.

Information for IEEE Authors. New York: American Institute of Electrical and Electronics Engineers, 1948; Supplement, 1965.

Instruction Manual for Preparing Research Papers, Minutes of Steering Committee Meetings, and Proposals. rev. ed. Chicago: Armour Research Foundation, 1958.

Instruction Manual Writing Guide. Doraville, Ga.: Scientific-Atlanta, n. d.

Instructions to Authors for the Preparation of Technical Papers. Metals Park, Ohio: American Society for Metals, 1962.

Manual for Authors of ASTM Papers. Philadelphia: American Society for Testing Materials, 1960.

Martin, M. J. Technical Writing and Speaking. Schenectady, N. Y.: General Electric Research Laboratory, 1952.

Middleswart, F. F. Instructions for the Preparation of Engineering
    Department Reports. rev. ed. Wilmington, Del.: E. I. du Pont,
    de Nemours & Company, 1953.
Miller, Elmo E. Designing Printed Instructional Materials: Content
    and Format. Arlington, Va.: Human Resources Organization,
    1975.
MLA Handbook for Writers of Research Papers, Theses, and Disser-
    tations. New York: Modern Language Association, 1977.
O'Rourk, John. Writing for the Reader. Maynard, Mass.: Digital
    Equipment Corporation, 1976.
Peters, Nade O. Laboratory Report Writer's Kit. St. Louis: Mc-
    Donnell Aircraft, 1963.
Petroleum Refiner Author's Handbook. Houston: Gulf Publishing,
    1958.
Practical Guide to Technical Communication. Burlington, Mass.:
    Raytheon Service Company, 1973.
Preparation of Engineering Reports. Baltimore, Md.: Glenn L.
    Martin, 1953.
Proposal Style Manual. Binghamton, N. Y.: Singer Company, 1976.
Publications Style Book. Thief River Falls, Minn.: Arctic Enter-
    prises, 1974.
Report and Letter Writing. Philadelphia: Sun Oil Company, 1955.
Report Standards. Dearborn, Mich.: Ford Motor Company, 1967.
Report Writer's Guide. 3rd ed. N. p.: Environmental Research
    Institute of Michigan, 1974.
Report Writing Tips. Pittsburgh: Westinghouse Research Labora-
    tories, 1974.
Ross, M. C. NRB Manual of Successful Business Letter Writing.
    Chicago: National Research Bureau, 1958.
Schaeffer, B. B., et al. Guide to Report Writing. 2nd ed. Niagara
    Falls, N. Y.: Olin Mathieson, 1958.
Schultz, Robert F. Preparing Technical Reports. Research Publica-
    tions GMR-427. Warren, Mich.: General Motors Corporation,
    Research Laboratories, 1964.
Siffin, Catherine F., et al. Author's Guide and Style Manual to
    Publications of the ERIC Clearinghouse on Reading. Preliminary
    ed. Bloomington: Indiana University, 1970.
Style Book and Editorial Manual. 5th ed. Chicago: Scientific Pub-
    lication Division, American Medical Association, 1971.
Style Manual. 6th ed. New York: Union Carbide and Carbon Cor-
    poration, 1954.
Style Manual. Rochester, N. Y.: Stomberg-Carlson, 1968.
Style Manual for Guidance in the Preparation of Papers for Journals
    Published by the American Institute of Physics and Its Member
    Societies. rev. ed. New York: American Institute of Physics,
    1977.
Style Manual for Technical Writers. Schenectady, N. Y.: General
    Electric Company Research Laboratory, 1963.
Swanson, Ellen. Mathematics into Type. Providence, R. I.: Amer-
    ican Mathematical Society, 1971.
Technical Publications Style Guide. Port Washington, Wis.: Sim-
    plicity Manufacturing Company, 1975.
Technical Report Manual. Detroit, Mich.: Engineering Division,
    Chrysler Corporation, 1955.

Technical Report Writing Procedure.   East Hartford, Conn.:   Pratt
   & Whitney Aircraft, 1955.
Technical Writing.   rpt. Washington, D.C.: Society for Technical
   Communication, n.d.   (Written for General Motors Research Lab-
   oratory, 1941.)
Van Buren, Robert, and Buchler, Mary Fran.   The Levels of Edit.
   Pasadena: Jet Propulsion Laboratory, California Institute of
   Technology, 1976.
Wallace, John D. , and Holding, J. Brewster.   Guide to Writing and
   Style.   rev. ed.   Columbus, Ohio: Battelle Memorial Institute
   Laboratories, 1966.
Writer's Handbook for John Deere Service Publications.   Moline,
   Ill.:   John Deere, 1972.

Government and Military Style Guides

Editorial Style Guide: A Publisher's Guide for Editors, Writers and
   Compositors of Scientific and Technical Reports.   China Lake,
   Calif.:   Naval Weapons Center Publications Division, 1975.
Effective Army Writing.   (Special Text 12-160.)   U.S. Army Adju-
   tant General School, July, 1964.
Effective Writing.   Washington, D.C.: Department of the Treasury,
   1975.
From Auditing to Editing.   Washington, D.C.: General Accounting
   Office, 1974.
General Services Administration.   Form and Guide Letters.   Wash-
   ington, D.C.:   U.S. Government Printing Office, 1973.
                                .   The 4-S Formula for Simplicity.   Washington,
   D.C.:   National Archives Records Service, 1960.
                                .   Plain Letters.   Washington, D.C.:   U.S. Govern-
   ment Printing Office, 1973.
Guide for Air Force Writing.   (AF Pamphlet 13-2.)   Washington,
   D.C.:   Department of the Air Force, 1973.
Guide for Preparation of Air Force Publications.   (AF Manual 5-1.)
   Washington, D.C.:   Department of the Air Force, 1955.
Guide for the Preparation of Geological Maps and Reports, 1965.
   rev. ed.   Ottawa, Canada: Department of Mines and Technical
   Surveys, 1965.
A Guide for the Development of Proposals, Progress, and Final Re-
   ports.   Frankfort: Bureau of Vocational Education, Kentucky
   State Department of Education, 1975.
Guidelines to Format Standards for Scientific and Technical Reports
   Prepared by or for the Federal Government.   rpt. Springfield,
   Va.:   Clearinghouse for Federal Scientific and Technical Informa-
   tion, 1968.
Holloway, A. H. , ed.   The Layout of Technical Reports.   Paris:
   NATO Advisory Group for Aeronautical Research and Develop-
   ment, 1956.
Improving Communication Through Effective Writing.   Washington,
   D.C.:   U.S. Department of Labor, 1957.
James, Frank W.   Job Performance Aid Methods (For Job Guide
   Manuals and Other Formats).   Dayton, Ohio: Air Force Logis-
   tics Command, Wright-Patterson Air Force Base, 1975.

Katzoff, S.   Clarity in Technical Reporting.   (NASA SP-7010.) Wash-
ington, D. C. :   U. S. Government Printing Office, 1964.
Klein, David.   The Army Writer.   4th ed.   Harrisburg, Pa. :   Mili-
tary Service Publishing Company, 1954.
Manual for NACA Editors.   Washington, D. C. :   National Advisory
Committee for Aeronautics, 1952.
Manual for NASA Editors.   Washington, D. C. :   National Advisory
Committee for Aeronautics, 1952.
ORNL Style Guide.   Oak Ridge, Tenn. :   Oak Ridge National Labora-
tory, 1974.
Preparation of Technical Publications.   Indianapolis: U. S. Naval
Aviation Facility, 1966.
Preparing and Processing Written Communications.   Manual 10-1.
Washington, D. C. :   Department of the Air Force, 1965.
Raaen, Helen P.   General Writing: Information Packet for the En-
vironment Impact Project.   Oak Ridge, Tenn. :   Oak Ridge Na-
tional Laboratory, 1975.
Remer, Ira, and Ziak, W. J.   Editing Guide.   Marshall Space Cen-
ter, Ala. :   NASA Publications, 1970.
Remer, Ira, and Ziak, W. J.   Writing Guide.   Marshall Space Cen-
ter, Ala. :   NASA Publications, 1967.
Siegel, Arthur I. , et al.   Techniques for Making Written Material
More Readable/Comprehensible.   Lowry Air Force Base, Colo. :
Air Force Human Resources Laboratory, 1974.
Smith, Frank R. , et al.   A Guide to the Formal Report.   Dayton,
Ohio: Air Force Institute of Technology, 1962.
Style Manual.   Ayer, Mass. :   U. S. Army Security Agency, Training
Center and School, 1966.
U. S. Air Force.   Report Writing Guide.   rev. ed.   Tullahoma,
Tenn. :   Arnold Engineering Development Center, 1958.
U. S. Army.   Style Manual.   (ST 32-4000.)   United States Army Se-
curity Agency Training Center and School, 1966.
U. S. Army Corps of Engineers.   Technical Report Writing.   Wash-
ington, D. C. :   Civilian Personnel Division, Office of the Chief of
Engineers, Department of the Army, 1955.
U. S. Department of the Air Force.   Military Standard MIL-STD-
847 (USAF):   Preparation of Technical Reports.   Washington,
D. C. :   U. S. Government Printing Office, 1966.
U. S. Department of Defense.   Defense Standardization Manual 4120. 3M:
Standardization Policies, Procedures, and Instructions.   Washing-
ton, D. C. :   U. S. Government Printing Office, 1966.
U. S. Department of the Interior.   Suggestions to Authors of the Re-
ports of the United States Geological Survey.   5th ed.   Washing-
ton, D. C. :   U. S. Government Printing Office, 1958.
U. S. Government Printing Office Style Manual.   Washington, D. C. :
U. S. Government Printing Office, 1973.
U. S. National Aeronautics and Space Administration Scientific and
Technical Information Divisions.   NASA Publications Manual.
Washington, D. C. :   U. S. Government Printing Office, 1964.
U. S. Soil Conservation Service.   Computation Standards for Engi-
neering Work.   Portland, Ore. :   Regional Technical Service Cen-
ter, Portland Design Section, 1969.
Van Hagen, Charles E.   Reviewing the Technical Report.   China
Lake, Calif. :   West Coast Navy Laboratories, 1959.

Graphics

Bain, Eric K. The Theory and Practice of Typographic Design.
New York: Hastings House, 1970.
Beakley, George C., Jr., and Autore, Donald D. Graphics for
Design and Visualization. New York: Macmillan, 1973.
Cameraready. Pasadena, Calif.: Cameraready Corporation, 1973.
Coffin, Harry Bigelow. Layout File for Printers and Advertisers.
(Direct Mail Advertising Association Research Report.) New York:
Moore, 1957.
Dicerto, J. J. Planning and Preparing Data-Flow Diagrams. New
York: Hayden, 1964.
Fetter, W. Computer Graphics in Communication. New York:
McGraw-Hill, 1965.
French, T. E., and Vierck, C. J. Graphic Science and Design.
3rd ed. New York: McGraw-Hill, 1970.
Garland, Ken. Graphics Handbook. New York: Van Nostrand Rein-
hold, 1966.
Gibby, J. C. Technical Illustration. 3rd ed. Chicago: American
Technical Society, 1969.
Giesecke, Frederick E., et al. Technical Drawing. 5th ed. New
York: Macmillan, 1967.
Glossary of Graphics and Technical Art Terms. Washington, D.C.:
Society of Technical Communication. n. d.
Guidry, Nelson P., and Frye, Kenneth B. Graphic Communication
in Science: A Guide to Format, Techniques, and Tools. Wash-
ington, D.C.: National Science Teachers Association, 1968.
Haemer, K. W. Making the Most of Charts. AMA Bulletin No. 28.
New York: American Telephone and Telegraph Company, 1960.
Haskell, A. C. How to Chart. Norwood, Mass.: Codex, 1947.
Hicks, G. A. Modern Technical Drawing. 2 vols. New York:
Pergamon, 1967-68.
Hill, P. H. Short Course in Engineering Graphics. New York:
Macmillan, 1964.
Hoelscher, R. P.; Springer, C. H.; and Dobrovolny, J. S. Graph-
ics for Engineers. New York: Wiley, 1968.
Karch, R. R. Graphic Arts Procedures. 3rd ed. Chicago: Amer-
ican Technical Society, 1965.
Levens, A. S. Graphics in Engineering and Science. New York:
Wiley, 1954.
Lutz, R. R. Graphic Presentation Simplified. New York: Funk
and Wagnalls, 1949.
Magnon, George. Using Technical Art: An Industry Guide. New
York: Wiley, 1970.
Modley, R., and Lowenstein, D. Pictographs and Graphs: How to
Make and Use Them. New York: Harper and Row, 1952.
Morris, George E. Technical Illustrating. Englewood Cliffs, N. J.:
Prentice-Hall Publishing Company, 1975.
Nelms, H. Thinking with a Pencil. New York: Barnes and Noble,
1964.
Pocket Pal: A Graphic Arts Production Handbook. 11th ed. New
York: International Paper Company, 1974.
Poulton, Eustace C. Effects of Printing Types and Formats on the

Comprehension of Scientific Journals.   Cambridge, England: Cam-
bridge University Press, 1959.
Pratt, K. C. House Magazine Layout.   Hamilton, Ohio: Champion
Paper Fibre, 1952.
Rogers, A. C. Graphic Charts Handbook.   Washington, D. C.: Pub-
lic Affairs Press, 1961.
Schmid, Calvin F., and Schmid, Stanton E. A Handbook of Graphic
Presentation. 2nd ed. New York: Wiley, 1979.
Stein, P. Graphical Analysis: Understanding Graphs and Curves in
Technology. New York: Hayden, 1964.
Stevenson, G. Graphic Arts Handbook and Products Manual.   Tor-
rance, Calif.: Pen and Press, 1960.
Stevenson, G. A. Graphic Arts Encyclopedia. New York: McGraw-
Hill, 1968.
Thomas, T. A. Technical Illustration. 2nd ed. New York: McGraw-
Hill, 1968.
Wiley, J. Barren. Communication for Modern Management.   Elm-
hurst, Ill.: Business Press, 1966.
Wood, Phyllis. Scientific Illustration. New York: Van Nostrand
Reinhold, 1979.

Publishing

Author's Guide for Preparing Manuscripts and Handling Proof. New
York: Wiley, 1950.
Bernstein, Theodore M. Headlines and Deadlines: A Manual for
Copy-Editors. New York: Columbia University Press, 1961.
Brookes, B. C., ed. Editorial Practice in Libraries. London:
ASLIB, 1961.
Butcher, Judith. Copy-Editing: The Cambridge Handbook. New
York: Cambridge University Press, 1975.
Chaundy, T. W.; Barrett, P. R.; and Batey, Charles. The Printing
of Mathematics. London: Oxford University Press, 1954.
Deaton, John G. Markets for the Medical Author. St. Louis: War-
ren Green II, 1971.
Gill, Robert S. The Author-Publisher-Printer Complex. 3rd ed.
Baltimore: Williams and Wilkins, 1958.
Heller, J. Printmaking Today. New York: Holt, 1958.
Hill, Mary, and Cochran, Wendell. Into Print: A Practical Guide
to Writing, Illustrating, and Publishing. Los Altos, Calif.:
Kaufmann, 1977.
Johnson, Donald F. Copyright Handbook. New York: Bowker, 1978.
Lasky, Joseph. Proofreading and Copy-Preparation. New York:
Mentor, 1954.
McNaughton, Harry H. Proofreading and Copyediting: A Practical
Guide to Style for the 1970's. New York: Hastings House, 1973.
A Manual of Style. 12th ed. Chicago: University of Chicago Press,
1969.
Melcher, Daniel, and Larrick, Nancy. Printing and Promotion
Handbook. 3rd ed. New York: McGraw-Hill, 1966.
New York Times Style Book for Writers and Editors. New York:
McGraw-Hill, 1962.

Rose, Lisle A.; Henter, Elmer F.; and Foster, George R.   Preparing Technical Material for Publication: A Manual for Authors
of College and Station Publications.   Urbana: University of Illinois Press, 1952.

Seltzer, Leon E.   Exemptions and Fair Use in Copyright: The Ex-
clusive Rights Tensions in the 1976 Copyright Act.   Cambridge,
Mass.: Harvard University, 1978.

Skillin, M. E., and Gay, R. M., eds.   Words into Type.   3rd ed.
New York: Prentice-Hall, 1974.

Susskind, C.   Dictionary of Style for Typewritten Technical Reports
and Manuscripts for Publication.   San Francisco: San Francisco
Press, 1960.

Turnbull, Arthur T., and Baird, Russell N.   The Graphics of Communication: Typography, Layout, Design.   3rd ed.   New York:
Holt, Rinehart and Winston, 1975.

Writer's Market '80.   Cincinnati: Writer's Digest, 1980.   Issued
annually.

Oral Communication

Anastasi, Thomas E., Jr.   Communicating for Results.   Menlo
Park, Calif.: Cummings, 1972.

Armstrong, Harold L., and Perigoe, J. Rae.   Getting Your Point
Across.   Mono. 1.   Toronto: Federation of Canadian Personnel
Associations, 1967.

Clyne, J. F., et al.   Business Speaking.   New York: Oxford University Press, 1956.

Dietrich, J. E., and Brooks, Keith.   Practical Speaking for the
Technical Man.   Englewood Cliffs, N.J.: Prentice-Hall, n.d.

Dyer, F. C.   Executive's Guide to Effective Speaking.   New York:
Prentice-Hall, 1962.

Frank, Ted, and Ray, David.   Basic Business and Professional
Speech Communication.   Englewood Cliffs, N.J.: Prentice-Hall,
1979.

Hand, Harry E.   Effective Speaking for the Technical Man: Practical Views and Comments.   New York: Van Nostrand Reinhold,
1969.

Hays, Robert.   Practically Speaking--In Business, Industry, and
Government.   Reading, Mass.: Addison-Wesley, 1969.

Howell, W. S., and Barmann, E. G.   Presentational Speaking for
Business and the Professions.   New York: Harper and Row, 1971.

Mambert, W. A.   Effective Presentations: A Short Course for Professionals.   New York: Wiley, 1977.

_____.   The Elements of Effective Communication.   Wash-
ington, D.C.: Acropolis, 1971.

_____.   Presenting Technical Ideas: A Guide to Audience.
New York: Wiley, 1968.

Manko, Howard H.   Effective Technical Speeches and Sessions: A
Guide for Speakers and Program Chairmen.   New York: McGraw-
Hill, 1969.

Michulka, Jean H.   Let's Talk Business.   Cincinnati: South-Western,
1978.

Miken, Ralph A. Speaking for Results: A Guide for the Business and Professional Speaker. Boston: Houghton Mifflin, 1958.

Morrisey, George L. Effective Business and Technical Presentation: Managing Your Presentations by Objectives and Results. 2nd ed. Reading, Mass.: Addison-Wesley, 1975.

Nadeau, Ray E., and Muchmore, John M. Speech Communication: A Career-Education Approach. Reading, Mass.: Addison-Wesley, 1979.

Phillips, D. C. Oral Communication in Business. New York: McGraw-Hill, 1955.

Powell, J. Lewis. Executive Speaking: An Acquired Skill. New York: BNA Education Systems, 1980.

Price, Stephen S. Business Ideas: How to Create and Present Them. New York: Harper and Row, 1967.

Sanford, W. P., and Yeager, W. H. Effective Business Speech. New York: McGraw-Hill, 1960.

Tacey, William S. Business and Professional Speaking. 2nd ed. Dubuque, Iowa: Brown, 1975.

Vardaman, George T. Effective Communication of Ideas. New York: Van Nostrand Reinhold, 1970.

Vogel, Robert A., and Brooks, William D. Business Communication. Brooks/Vogel Series in Speech Communication. Menlo Park, Calif.: Cummings, 1977.

Weiss, Harold, and McGrath, J. B., Jr. Technically Speaking. New York: McGraw-Hill, 1963.

Wilcox, Roger P. Oral Reporting in Business and Industry. Englewood Cliffs, N.J.: Prentice-Hall, 1967.

Zelko, H. P., and Dance, F. E. X. Business and Professional Speech Communication. New York: Holt, Rinehart, and Winston, 1965.

Zollinger, Robert Milton; Pace, William G.; and Kienzle, George J. A Practical Outline for Preparing Medical Talks and Papers. New York: Macmillan, 1961.

Style, Language, and Readability

Bernstein, Theodore M. The Careful Writer. Baltimore: Penguin, 1975.

_____. Do's Don'ts and Maybes of English Usage. New York: Times Books, 1977.

_____. Miss Thistlebottom's Hobgoblins. New York: Farrar Straus, 1971.

British Council. English for Academic Study with Special Reference to Science and Technology: Problems and Perspectives. London: English Teaching Information Centre, 1975.

Chall, J. S. Readability--An Appraisal of Research and Application. Columbus: Ohio State University Press, 1958.

Close, R. A. The English We Use for Science. London: Longman, 1965.

Colwell, Carter C., and Know, James H. What's the Usage: The Writer's Guide to English Grammar and Rhetoric. Englewood Cliffs, N.J.: Prentice-Hall, 1972.

Condon, John C. Semantics and Communication. New York: Macmillan, 1966.

Copperud, Roy H. American Usage: The Consensus. New York: Van Nostrand Reinhold, 1970.

_____. A Dictionary of Usage and Style. New York: Van Nostrand Reinhold, 1970.

Corbett, Edward P. The Little English Handbook: Choice and Conventions. New York: Wiley, 1973.

Durham, John, and Zall, Paul. Plain Style. New York: McGraw-Hill, 1967.

Evans, Bergen, and Evans, Cornelia. A Dictionary of Contemporary American Usage. New York: Random House, 1957.

Flesch, Rudolf F. The ABC of Style: A Guide to Plain English. New York: Harper and Row, 1964.

_____. The Art of Plain Talk. New York: Harper and Brothers, 1946.

_____. The Art of Readable Writing: 25th Anniversary Edition. Scranton, Pa.: Harper and Row, 1974.

_____. How to Be Brief. New York: Harper and Brothers, 1962.

_____. How to Make Sense. New York: Harper and Brothers, 1954.

_____. How to Say What You Mean in Plain English. New York: Barnes and Noble, 1974.

_____. How to Test Readability. New York: Harper and Brothers, 1951.

_____. Say What You Mean. New York: Harper and Row, 1972.

Follett, Wilson. Modern American Usage. New York: Hill and Wang, 1966.

Fowler, H. W. A Dictionary of Modern English Usage. rev. by Sir Ernest Gowers. New York: Oxford University Press, 1965.

Gopnik, Myrna. Linguistic Structures in Scientific Texts. (Janua Linguarum, Series Minor, 129.) The Hague: Mouton, 1972.

Gowers, Sir Ernest. The Complete Plain Words. Baltimore: Penguin, 1975.

Harbaugh, Frederick W. Think It Clearly, Make It Tell--With Information Impact. North Quincy, Mass.: Christopher, 1978.

Hayakawa, S. I. Language in Thought and Action. 3rd ed. New York: Harcourt, Brace and World, 1978.

Herbert, A. J. The Structure of Technical English. New York: Longman, 1975.

Hook, Julius. Guide to Good Writing. New York: Ronald, 1962.

Huddleston, Rodney D. The Sentence in Written English: A Syntactic Study Based on an Analysis of Scientific Texts. Cambridge Studies in Linguistics 3. London: Cambridge University Press, 1971.

Klare, George R. The Measurement of Readability. Ames: Iowa State University Press, 1963.

Klare, George R., and Buck, B. Know Your Reader: The Scientific Approach to Readability. New York: Hermitage, 1954.

Lambuth, David, et al. The Golden Book on Writing. 2nd ed. New York: Viking, 1964.

Mager, N. H., and Mager, S. K. Encyclopedia of English Usage. Englewood Cliffs, N.J.: Prentice-Hall, 1974.

Morris, William, and Morris, Mary, Eds. Harper Dictionary of
    Contemporary Usage. New York: Harper and Row, 1975.
Nicholson, Margaret. American English Usage. London: Oxford
    University Press, 1957.
Nickles, Harry G. The Dictionary of Do's and Don'ts: A Guide for
    Writers and Speakers. New York: McGraw-Hill, 1974.
Partridge, Eric. Concise Usage and Abusage. New York: Philo-
    sophical Library, . 1955.
Perrin, Porter G., and Ebbitt, W. R. Writer's Guide and Index to
    English. 5th ed. Chicago: Scott, Foresman, 1972.
Quiller-Couch, Sir Arthur. On the Art of Writing. New York:
    Putnam's, 1916.
Savory, T. H. The Language of Science--Its Growth, Character
    and Usage. London: Deutsch, 1953.
Stone, W., and Bell, J. G. Prose Style: A Handbook for Writers.
    New York: McGraw-Hill, 1968.
Strunk, William, Jr., and White, E. B. The Elements of Style.
    3rd ed. New York: Macmillan, 1979.
Turabian, Kate L. A Manual for Writers of Term Papers, Theses,
    and Dissertations. 3rd ed. Chicago: University of Chicago
    Press, 1967.
Wood, Frederick T. Current English Usage. London: Macmillan,
    1962.

The following titles were discovered while this bibliography was in production.

A1 Bosticco, Mary. <u>Personal Letters for Businessmen</u>. 2nd ed.
      London: Business Publications, 1966. 330p.
This British book contains 13 chapters dealing with personal business letters. Chapter 1, "Getting Off to a Good Start," discusses effective style and punctuation and gives correct forms of address for royalty, high legal officials, diplomats, clergy, and military officers.
Other chapters cover such topics as "Letters of Introduction," "Letters of Recommendation," "Thank You Letters," and "Requests."
Chapter 13, "Miscellaneous," includes letters of regret and apology and letters by a secretary on behalf of an absent supervisor. Each chapter contains a page of advice followed by numerous sample letters.

A2 Bromage, Mary C. <u>Writing Audit Reports</u>. New York: McGraw-
      Hill, 1979. Index. Bib. 193p.
The author asks all "who pursue the unending search for precise and persuasive communication on paper" to read this book as "policies and procedures" rather than rules. The 11 chapters discuss such topics as "Choosing Words," "Streamlining Sentences," "Putting Paragraphs Together," "Designing Format," and "Strengthening Content."
Chapter 6 covers "Considering Auditors as Authors" and discusses how auditors' personal, professional, and organizational characteristics show up on paper. Auditors are "exact, critical but constructive" as they "incorporate formally the observations and conclusions from a professional analysis."

A3 Coleron, Henry C., and Furt, F. Allen. <u>How to Conduct Army</u>
      <u>Correspondence</u>. New York: Harper, 1943. 119p.
This book was prepared for both "the officer who dictates and the enlisted man or civilian clerk who types the messages." The 20 chapters cover such topics as "Military Letters," "Progressive Steps in the Construction of a Military Letter," "Secret, Confidential,

and Restricted Communications or Documents, " and "Dictating the
Letter. " Much of the focus is on format. The Appendix provides
an explanation of the organization of Army Regulations. Also in the
Appendix is the index to Army Regulations 340-15 (August 21, 1942),
Correspondence--How Conducted. A list of abbreviations and symbols and an organizational chart with explanation of duties of Army
officers are included.

A4 Craig, Robert J. , and Yeats, Harry W. The Complete Guide
        for Writing Technical Articles. Norcross, Ga.: American
        Institute of Industrial Engineers, 1976. 25p.
This work is aimed at "those professionals who have many good ideas
to offer, but who have been reluctant to try their hand at writing. "
Chapter 1, "Why Write Technical Articles?, " discusses motivations
for technical publication, defines the technical article, and explains
the preliminary steps of selecting a subject and querying a journal.
Chapter 2, "The Ordeal (Writing a Technical Article), " explains the
writing steps, organization, style, and revision. Chapter 3, "Mechanics of Preparing the Article, " discusses the use of graphics and
documentation. Chapter 4, "Final Steps in Processing Your Manuscript, " covers submission as well as proofreader's marks and how
to read galleys. Chapter 5 is a list of "Publications, Addresses
and Other Pertinent Data" covering 14 subjects, such as "Computers
and Data Processing"; "Health, Hospitals, and Services"; and "Quality Control. "

A5 Crawford, Jack and Kielsmeier, Cathy. Proposal Writing.
        Corvallis, Ore.: Continuing Education, 1970. 169p.
This book is divided into four parts. Part I, "Manual, " is a 19-page
discussion of objectives, audience, statement of the problem, procedures, measurement, and financial matters. Part II, "Appendices, "
consists of Appendices A-I. Appendix A is a two-item bibliography
of sources. Appendices B-E are short lists of sources on such
points as procedures, budgets, and funding. Appendix F shows the
results of a 1959 study on why proposals fail. Appendices G-I are
examples of an experimental research proposal, a developmental
project proposal, and an informal developmental proposal. Part III,
"Workbook, " concentrates on illustrating the components of a proposal. Examples from actual proposals are included. Part IV, "Evaluating Proposals, " provides samples of forms used by HEW to evaluate proposals and two proposals submitted to HEW Regional offices.

A6 Culliton, James W. Writing Business Cases. Boston: Gradu-
        ate School of Business Administration, Harvard University,
        1946. 30p.
This work was written primarily for "research assistants and others
engaged in preparing teaching material for use at the Harvard Business School" and is published "in the hope that it may shed some
light on a little-discussed aspect of the case method of instruction. "
Chapter 1, "Case Writing--Why and By Whom, " explains that professors must rely on assistants to do case research because it is
so time-consuming. Chapter 2, "The Case Collection Process--The
Interview, " gives the steps of preparing for the interview and proper

questioning techniques.   Chapter 3, "The Case Collection Process--
After The Interview, " deals with keeping records and getting company
releases to use information.   Chapter 4, "Techniques of Case Writing, " discusses such points as organization, stating the issue, and
graphics.   Chapter 5, "The Effective Use of Words, " covers such
items as reader, diction, conciseness, and paragraphs.   The author
comments that one of the chief values of the case method is that it
"forces teachers to observe, study, and understand the realities of
a situation" and so overcome "professional tendencies to make ex
cathedra pronouncements. "

A7   Drummond, Gordon.   English for International Business.   Lon-
         don: Harrap, 1970.   159p.
This book for nonnative speakers of English emphasizes phraseology
and layout of the business letter.   Part I, "Business Letters and
Telephone Calls in English, " has 16 chapters showing sample letters
and giving dialogue for telephone calls on such topics as accounts
and invoicing, insurance, hotel bookings, and inquiries.   Most samples are in the British style.   Part II, "General Information, " contains Sections A-F.   Section A deals with telecommunications.   Section B gives samples of letters in the American style including layout and spelling.   Section C is a glossary comparing British and
American vocabulary and spelling.   Section D explains British and
American money, and Section E shows weights and measures from
British to metric.   Section F is a list of common abbreviations and
signs.

A8   Fitting, Ralph U.   Report Writing.   New York: Ronald, 1924.
         Index.   100p.
This book deals with both business and technical reports.   Chapter
1, "The Nature of a Report, " discusses the types of reports and
their uses.   Chapter 2 covers the "Period Report" and discusses
clarity, selection of detail, and format.   Examples from the 1919
annual report of the American Cyanamid Company and the 1918 annual report of the Atchison, Topeka and Santa Fe Railroad System
are included.   Chapters 3-6 cover the "Examination Report. "   The
author concentrates on the four leading types: financial, engineering,
organization or efficiency, and marketing reports.   Topics include
discussions of the objectives, sources of materials, scope, and the
writer of such reports.   In Chapter 4 the author explains investigation and discusses the investigator's attitude, thoroughness, and experience.   The discussion of fieldwork covers interviews, questionnaires, note-taking, sampling, and personal observation.   Examples
of the steps are based on the investigation of a project to electrify
a "certain western railroad. "   Chapter 6 deals with organization.
Sample outlines for a financial report, a public utility report, a
mining report, and an industrial report are included.   The author
says engineering reports are too varied to provide sample outlines.
Chapter 7 covers "Writing Up the Report" and discusses readability,
tone, and visuals.

A9   Hollis, Joseph W. , and Donn, Patsy A.   Psychological Report
         Writing: Theory and Application.   Muncie, Ind.:   Accel-
         erated Development, 1973.   Index.   Bib.   273p.

This book, set in typewriter font, is divided into four parts. Part
I, "Purposes, Methods, and Philosophy," contains three chapters
that focus on the purposes of reports from three viewpoints: the
writer's, the user's, and the administrative. Also covered are methods of gathering data, sampling, and the psychological philosophies
expressed in reports. Part II, "Confidentiality and Security," has
three chapters discussing privacy, release of information, and data
control. Part III, "Writing Style and Error," contains two chapters
giving tips on clarity and format, as well as warning about common
errors, such as gobbledygook, wordiness, and the dangers of being
misunderstood. Part IV, "Outlines, Samples and Forms," contains
samples divided into six categories. Throughout the book cartoons
illustrate concepts.

A10 Horn, Ernest, and Peterson, Thelma. The Basic Vocabulary
        of Business Letters. New York: Gregg, 1943. 236p.
This work is based on Horn's earlier studies of the frequency with
which words were used in all correspondence. For this book only
business correspondence was examined. Letters were used from
26 classes of businesses, such as florists, law firms, and banks.
The average number of firms in each class was six. The book lists
some 8,000 words alphabetically. Each word carries with it numbers indicating its relative frequency. For example, the word consider is ranked in the first half of the first thousand words. It was
found 414 times in 23 classes of businesses. It was given a weighted
total of 2,070, a figure reached by multiplying 414 by 5, which is
the nearest integer of the square root of 23. Appendix A contains
5,226 most-used words listed in order of frequency. Appendix B
lists abbreviations and the number of times they appeared in the
sampling. For example, Co. appeared 286 times with a weighted
total of 1,144. Appendix C lists fractions and numbers and their
frequencies.

A11 Ireland, Stanley. How to Prepare Proposals That "Sell."
        Chicago: Dartnell, 1967. 140p.
This book, set in typewriter font, has six chapters. Chapter 1,
"Introduction," discusses the salesmanship aspect of proposals and
the competition. Chapter 2, "The Marketing Decision: To Bid or
Not to Bid," includes a checklist for a marketing evaluation of a
proposal and a chart showing the steps to winning a contract. Chapter 3, "Plans and Preparation," covers such points as content, responsibility for preparation, cost, and organization. Chapter 4,
"Elements and Factors of an Effective Proposal," focuses on audience, front matter, design, and sections of the proposal. Chapter
5, "Evaluation and Analysis," explains criteria for analysis and includes charts for evaluation of proposals. Chapter 6 contains a
checklist for proposal evaluation, a full sample of a proposal of a
manufacturer that includes the management proposal and the technical proposal, and examples of cover letters. Within the chapters
are further illustrations of proposal elements.

A12 Martin, William T. Writing Psychological Reports. Spring-
        field, Ill.: Thomas, 1972. Index. Bib. 184p.
This book is aimed at the "novice writer who has never written a

psychological report. " The 15 chapters cover such topics as "Purpose and Content of the Report, " "Organization of the Report, " and "Language of the Report. " Chapters 6-11 cover specific sections of the report, such as "The Diagnosis, " "The Personality Evaluation, " and "The Treatment Summary. " Numerous samples of reports are included. Chapter 14 discusses the "Legal Implications of Communication, Testimony, and Reporting. " Sample forms for release of information or medical records are included.

A13   Papers Presented at the All-Ohio Conference on Technical Communications.   Cincinnati: Society of Technical Writers and Editors, 1959.   71p.
This conference, held September 11-12, 1959, was sponsored by the Dayton-Miami Valley Chapter of the publishing society. Session I, "Principles, " includes such papers as "The Semantics of Technical Writing, " and "Logic and Methods in Written Technical Communications. " Session II, "Technical Writing in Industry, " features such papers as "The Case for In-House Technical Writing" and "The Case for Contracting Technical Writing. " Session III covers "Technical Writing in Government" and includes "Technical Writing in the Air Material Command. " Among the authors of papers are the assistant director of the Battelle Memorial Institute; a captain at the Institute of Technology Air University, USAF; and the manager of product support publications of General Electric Company.

A14   Ridley, Clarence E. , and Simon, Herbert A.   Specifications for the Annual Municipal Report.   Chicago: International City Managers' Assoc. , 1939.   Bib.   59p.
This work, set in typewriter font, has three chapters covering content and presentation, preparation of copy and publication, and layout and design.   Illustrations of actual municipal reports from such cities as Saginaw, Michigan; Rochester, New York; San Jose, California; and Portland, Oregon, are included.   The appendix is a one-page bibliography.

A15   Robinson, William L.   R&D Proposal Preparation Guide.   N. p. : Queensmith Associates, 1962.   37p.
This work, set in typewriter font, has eight chapters covering such topics as "General Instructions for Proposal Preparation, " "Production of Proposals, " and "Government Evaluation of Proposals. " Chapter 4, "The Technical Proposal, " discusses content, organization, the draft, and the sections of the proposal.   The focus of this work is on R&D proposals to the government and to industrial firms doing business with the government.

A16   Tilghman, William S.   Your Technical Proposal.   Washington, D. C. : DATA Publications, n. d.   Bib.   29p.
This work is divided into an introduction, five sections, and appendixes, each part on different-colored paper.   The pages are cut at graduated lengths to allow the reader to flip open at once to a special section.   Topics covered included "Proposal Summary, " "The Problem, " and "Proposal Solution. " The four appendixes are on

unnumbered pages and contain a bibliography, a discussion of the duties of the proposal manager, tips on résumé format, and discussion of the "boilerplate" material in a proposal (company description and history).

Works listed with the prefix "A" (A1, A2, A3, etc.) appear in the Addendum.

Except for "Early Books" categories, this subject index contains only books that are annotated. For additional topics, see also "Related Works," pp. 197-207, and the Title Index.

Technical Writing (Professional Guides) 33, 60, 66, 67, 147, 158, 302, 304, 359, 360, 393, 413, 492, 508, 545, 754, 764

Textbooks, General (See specific categories, such as Engineering Writing and Scientific/Science Writing)

Textbooks, Business Reports 13, 15, 37, 62, 101, 106, 235, 268, 346, 442, 464, 476, 635, 707, 765

Textbooks, Business Writing 4, 10, 17, 30, 50, 52, 57, 78, 80, 86, 90, 91, 92, 94, 97, 102, 105, 112, 122, 152, 183, 199, 211, 212, 223, 238, 240, 242, 280, 281, 284, 292, 323, 331, 342, 345, 351, 353, 364, 366, 375, 376, 386, 387, 388, 400, 407, 440, 445, 449, 457, 460, 461, 462, 472, 480, 481, 489, 498, 501, 514, 517, 527, 530, 531, 564, 567, 595, 597, 617, 621, 650, 651, 652, 658, 661, 682, 693, 704, 706, 709, 715, 733, 742, 743, 745, 756, 757, 760, 770, 788, 798, 801, 809, 812, 818

Textbooks, High School 6, 24, 81, 116, 219, 254, 285, 287, 331, 495, 537, 717, 736, 821

Textbooks, Technical Reports 15, 36, 70, 115, 176, 198, 313, 379, 522, 572, 642, 673, 688, 726, 735, 763, 765, 766, 795, 825, 828

Textbooks, Technical Writing 5, 11, 16, 45, 49, 74, 173, 194, 197, 264, 265, 281, 301, 310, 334, 341, 347, 354, 368, 372, 408, 414, 447, 469, 486, 515, 539, 542, 543, 551, 594, 605, 633, 667, 699, 733, 749, 750, 755, 778, 793, 797, 807, 816, 830

Textbooks, Vocational/Occupational 47, 48, 108, 109, 123, 165, 338, 382, 420, 452, 490, 529, 582, 608, 614, 618, 664, 711, 737

Translation, Technical Writing 746 (See also English as a Second Language and Model Letters in Translation)

Usage/Spelling Guides 145, 217, 451

Vocational Writing (See Textbooks, Vocational/Occupational)

Workbooks 47, 51, 79, 134, 165, 216, 217, 268, 285, 333, 351, 353, 386, 387, 388, 419, 440, 449, 451, 490, 500, 529, 547, 638, 664, 696, 697, 742 (See also Textbooks categories)